Humans-with-Media and the Reorganization of Mathematical Thinking

Mathematics Education Library
VOLUME 39

The titles published in this series are listed at the end of this volume.

Marcelo C. Borba
Mónica E. Villarreal
(Authors)

Humans-with-Media and the Reorganization of Mathematical Thinking

Information and Communication Technologies, Modeling, Visualization and Experimentation

Preface by Ubiratan D'Ambrosio
Afterword by Ole Skovsmose

 Springer

Library of Congress Cataloging-in-Publication Data

Borba, Marcelo C. (Marcelo Carvalho)
  Humans-with-media and the reorganization of mathematical thinking: information and
communication technologies, modeling, visualization, and experimentation / Marcelo C.
Borba, Mónica E. Villarreal; preface by Ubiratan D'Ambrosio; afterword by Ole Skovsmose.
    p. cm.—(Mathematics education library; v. 39)
  Includes bibliographical references and index.
  ISBN: 0-387-24263-5   eISBN: 0-387-24264-3
    1. Mathematics—Study and teaching—Methodology. 2. Communication in science.
  I. Villarreal, Mónica E. II. Title. III. Series.

QA11.2.B67 2005
510'.71—dc22   2004065301

Printed in the United States of America.

9 8 7 6 5 4 3 2   SPIN 11589884

springer.com

Dedication

This book is dedicated to

Anne Kepple

Contents

Dedication   v

Contributing Authors   xi

Preface by Ubiratan D'Ambrosio   xiii

Foreword   xvii

Acknowledgments   xxi

Why another book about technology and mathematics education?   1
    1. Introduction   1

Information technology, reorganization of thinking and
    humans-with-media   9
    1. The second industrial revolution and education   9
    2. Reorganization of thinking   11
    3. Intershaping relationship: stressing balance   15
    4. Humans and technology: a history of separation   17
    5. Media, humans and knowledge: possibilities of merging   21

Modeling as a pedagogical approach: resonance with new media   29
    1. Introduction   29
    2. Problem solving, problem posing and modeling   33
    3. Roots of modeling in Brazil   47
    4. Project work: its roots in Denmark   52

viii

5. Modeling and information and communication technology   54
6. Modeling and its limitations   59

Experimental-with-technology approach: resonance with modeling and
    multiple representations   63
1. Introduction   63
2. Experimentation in mathematics   65
3. Experimentation in mathematics education   71
4. Multiple representations and media   76

Visualization, mathematics education and computer environments   79
1. Visualization: some definitions   79
2. Visualization and media in mathematics   82
3. Visualization and media in mathematics education   88
4. Visualization and humans-with-media   97

Modeling and media in action   101
1. Introduction   101
2. Modeling in a mathematics course for biology majors   102
3. Modeling and humans-with-textbooks-Excel-paper-and-pencil
    collectives   105
4. Modeling when the Internet becomes an actor   109
5. Modeling, humans-with-paper-and-pencil and... potatoes   114
6. Science in action in the classroom and video clip culture   119

Experimentation, visualization and media in action   125
1. Introduction   125
2. Experimenting with parabolas: visual conjectures   126
3. Experimenting with conic sections: more visual conjectures   130
4. Experimenting with functions I: the AG-GA theorem   135
5. Experimenting with functions II: multiple representations and
    intermedia coordination   140
6. Construction of derivatives: a graphical approach   145
7. Tangent lines: visual and algebraic approaches   151
8. Visualization, media and the voice of the students   156
9. Visualization, experimentation and books   158
10. Experimentation, visualization and reorganization of thinking   165

Mathematics and Mathematics Education on-line   169
1. Humans-with-Internet and education   169
2. The nature of interaction in a distance education course   173
3. Chat and mathematics in the classroom   179
4. Research problems   184

Methodology: an interface between epistemology and procedures   187

Political dimensions of Information and Communication Technology   201

Afterword by Ole Skovsmose   211

References   217

Index   227

Contributing Authors

Marcelo C. Borba is a professor of the graduate program in Mathematics education and of the mathematics department of UNESP (State University of Sao Paulo), campus of Rio Claro, Brazil. He is a member of the editorial board of *Educational Studies in Mathematics* and a consultant for several journals and funding agencies both in Brazil and abroad. He is the editor of BOLEMA. He wrote several books, book chapters and papers published in Portuguese and in English, and he is the editor of a collection of books in Brazil, which includes ten books to date.

Mónica E. Villarreal is a calculus professor at the Faculty of Agronomy of the University of Córdoba. She concluded her doctorate in mathematics education at UNESP, Rio Claro. She has supervised Masters students and has directed various research projects in Argentina. She is well published in Spanish, Portuguese and English. She is a consultant of BOLEMA, one of the most important mathematics education journals in Brazil, and of Revista de Educación Matemática, a journal in Argentina.

Preface

As we enter into the 21st Century, the presence of technology, particularly techno-science, in everyday life is overwhelming. Institutions in the modern world are affected by this presence.

Like Janus, the ancient Roman god whose double-faced head signified his knowledge of the present and the future, education has always been a two-faced enterprise. The past establishes goals and methods of Education, and the other face tries to capture the future and suggests and proposes new directions of thought and new styles of behavior for the generation which, in a few years, will take over both routines and societal innovation. History tells us that this face of Education has always been sensitive to emerging technologies.

Technologies of communication and information have been particularly influential in new directions of society, in particular of education. The transition from orality to writing marked a new role for the teacher. From the sole repository of accumulated knowledge, the teacher became a guide and interpreter of registered knowledge. The emergence of hardware, in the form of documents and books, initiated a companionship between teacher and hardware. It is also remarkable how the emergence of writing strengthened individual memory, contrary to the concerns of Thamus when Theuth explained to him the discovery of writing. The conservative king was afraid that the new invention would implant forgetfulness in the souls of men. Something similar occurred in Europe with the introduction of the technology of calculation of Indian and Arabic origins, which strengthened the analytic instruments of the philosophers of the late European Middle Age, thus paving the way for the Renaissance and Modern Age. We are now living new possibilities in our communicative and analytic capabilities, thanks to the powerful new technology of communication and information.

Marcelo Carvalho Borba and Mónica Villarreal embraced the Janus metaphor when they decided to write this book. They are able to review, critically, the most relevant current educational practices, which largely reflect our past, and to venture into the future, proposing new directions for education. The same care of critically regarding the past is present in their views of the future. Thus, this book does not get trapped by the marvels suggested by the new, amazing, technologies.

It is a fact that billions are spent in education worldwide. But they risk being lost if we insist on declining educational models and practices. This big loss is unbearable for most countries, where human resources, so necessary for their future, receive an obsolete, and in most cases, useless, education. Even the more prosperous economies are very much concerned with the downgrading of their education, in spite of enormous resources available. We all agree that technology, by itself, is not the guarantee of a good education. But it is undeniable that lack of technology may hinder progress in education. Borba and Villarreal point to the key issues related to this paradoxical situation, avoiding sameness.

The book has an exemplary organization. In 10 chapters, the authors examine all the issues raised by the emerging technologies which are relevant for mathematics education. The challenges to the educator, from the cognitive dimensions to the political issues, are all dealt with by the authors. Although the book has originated in Brazil, the concerns are common to both the less and the more prosperous economies.

The authors claim a *de facto* evolution of the species towards higher levels of humanity, in the sense of a species impregnated with respect, solidarity and team spirit. This is particularly noticeable when they focus on the interaction of humans and technology. Refusing a common concern that technology leads to lack of humanity, the authors, drawing from many examples from the history of culture, claim the opposite. Indeed, there has been an interaction between humans and the technology they have created, and the evolution of the human species results from this interaction, to the point of a true merging of technologies in everyday life and, remarkably, in the way we think and act. The authors examine these facts, pointing out, very convincingly, that it is the responsibility of education to guide this merging to the ultimate goal of humanity. This is absolutely necessary for the survival, with dignity, of civilization.

The trajectory to a species impregnated with respect, solidarity and team spirit meets with obstacles of a political nature. We may appeal, again, to the Janus metaphor when referring to the conservative opposition to the new, to a trend in Education to favor sameness. It is the strength of the ethos of a society that supports this face. On the other side, acquiescence allows the absorption of the new. Caution, necessary in every step of human action, should not hinder venturing into the new. This is another merit of this book. The authors, well aware of the need of caution, implemented, with all the required instruments of monitoring and evaluation, many innovative projects. Most of the projects of technological innovation in Mathematics Education, internationally recognized, received attention of the authors and were the subject of careful research. The description of the projects, accompanied by the results of their research and by very important remarks, will be extremely valuable for those wishing to innovate. A rich bibliography helps the mathematics educators in the process of carrying on their projects inspired by this book.

Borba and Villarreal have written an excellent book. Combining high scholarship with sound and careful methodology, they give to the reader, not only mathematics educators, a support for being innovative in entering the future.

Writing this preface was a most pleasant experience. It is always an honor to be invited by colleagues and friends to write a preface. But when the authors are former students, as in this case, the honor is multiplied. I am reassured that the good moments we had together were enriching to all of us.

Ubiratan D'Ambrosio
São Paulo, August 2004

Foreword

The ideas that have matured and given fruit in this book have their roots in diverse places, and here we would like to tell part of the story of the path leading to the book's completion.

In 1988, Marcelo Borba began his doctoral studies at Cornell University, U.S.A, becoming a member of the Mathematics Education Research Group, led by Jere Confrey. Since then, he has reflected on computers, their presence in education, and related epistemological issues, until arriving at the notion of *humans-with-media*, the backbone of this book, which represents an attempt to break the dichotomy between humans and technology, with relevant consequences for the classroom.

He returned to his native Brazil in 1993, where he started working at the State University of São Paulo (UNESP), one of the most important centers of mathematics education in Brazil and Latin America, and where he had earned his masters degree in 1987 conducting research on ethnomathematics. Former concerns about culture, social justice, and political dimensions of education - although never abandoned by him while studying abroad - gained new impulse and began to interact with some of the epistemological issues related to computers. He began making the connections between ethnomathematics, phenomenology, epistemology, technology and mathematics, and a project for a book started to develop. An outline of the book was presented to Kluwer and approved in 1996. Personal problems and other professional assignments made it impossible to finish the book at that time.

These problems, however, seem to have had positive consequences. Some of the ideas, especially the notion of humans-with-media, matured and became intertwined with new data from ongoing research. Several articles were published, mainly in Portuguese, and the project of the book was revised. Although Lévy continues to be the main reference for this work, with his notion of the thinking collective and his view of technology, the writings of other authors like Kerckhove and Castells helped to transform some of the main ideas. He has been bringing Levy's ideas to mathematics education for almost a decade, but in this book, he intertwines his ideas with examples from research in new ways.

Other changes took place, as well, especially when he invited Mónica Villarreal, from the University of Córdoba, Argentina, to help him with this endeavor. The original invitation was for her to help with the literature review and some specific chapters, however she quickly transformed into a co-author of the book, helping to clarify many of the ideas, and adding her touch to it. Villarreal came twice to Brazil to work on the book, in 2002 and 2003, sponsored by FAPESP, a funding agency of the State of São Paulo, Brazil. Prior to that, she obtained her doctoral degree in the Graduate Program of Mathematics Education from UNESP, Rio Claro, in 1999. It was during this period that she started to interact with Marcelo Borba as his advisee and member of their research group, GPIMEM. The constant dialogue, exchange, informal conversation, and the contact with new research perspectives and new authors, changed Mónica Villarreal's perspective regarding what it means to teach and learn mathematics and to conduct research in mathematics education. Upon her return to Argentina, she resumed her activities as researcher and professor at the University of Córdoba, initiating a new phase in her relationship with GPIMEM as an associated researcher.

Anne Kepple, a researcher in public health nutrition, made an enormous contribution to this book. Although her job was to edit the English written by a Brazilian and an Argentinean, she did much more than this. As someone who knows qualitative research very well, she helped us to balance the book by challenging claims and assertions made in earlier versions of it. She also suggested readings, paths to be followed in the investigations, and provided motivation when some obstacles seemed insurmountable. We would like to thank her in a very special way.

We also would like to thank Ole Skovsmose who motivated Marcelo Borba every chance he had to publish this book as he believed that the ideas regarding technology and mathematics education were original. We would also like to thank Ubiratan D'Ambrosio and Maria Bicudo who have strongly influenced us on issues regarding ethnomathematics and education in a dialogical perspective. The reader will see that parts of this book have been previously published in different forms and languages. The reader will always find complete references in the text to these publications.

Finally, it should be said that this book represents a consolidation of a way of conceptualizing research group, as Mónica Villarreal had been a regular member of the group, from 1995 through 1999, taking part in regular weekly meetings; and has since become an associated member of the group, which is a category created for members who have special projects, but do not participate regularly in the group, and are more likely to develop research within other research groups as well. Since 2000, we have worked together on-line and face-to-face, consolidating a fruitful thinking collective, and sharing research perspectives and academic discussions, intertwined with research conferences, dance sessions, and the very important, fun and indispensable conversations accompanied by good red wine.

Marcelo C. Borba and Mónica E. Villarreal

Acknowledgments

We would like to thank:

CNPq, a funding agency of the Brazilian government, that has sponsored projects by the first author of this book for the last eleven years (Grants: 520033/95-7, 471697/2003).

CAPES, another funding agency of the Brazilian government, who funded the doctoral work of the second author of this book from 1996 through 1999 (PROAP Program).

FAPESP, a funding agency of the State of São Paulo, that provided funding for the second author of this book to travel to UNESP, Rio Claro, SP, Brazil, in 2002 and 2003, so that this book could be concluded (Grants:2002/07290-0, 2003/03750-9).

All the students who were involved in our research.

The reviewers of this book and of the grant proposals we have received.

The Faculty of Agronomy of University of Córdoba which provided the leave of the second author of this book to develop research at UNESP.

The Graduate Program of Mathematics Education and the Mathematics Department at UNESP, Rio Claro, which provided the infrastructure for the development of this book.

All members and former members of our research group, GPIMEM, who influenced us directly in the development of this book - not only because of the mutual influence on our ideas, but because of their sharp criticism of earlier versions of some of the chapters. We would like to thank, in particular, Ana Paula Malheiros, Telma Gracias, Jonei Barbosa, Jussara Araújo, Nilce Scheffer, Francisco Benedetti, Rúbia Amaral, Audria Bovo, Fernanda Bonafini, Norma Alevatto, Miriam Penteado, Marcus Maltempi, Antonio Olímpio Junior, Ricardo Scucuglia, Ana Flávia Mussolini, Simone

Lírio, Adriana Richt, Simone Gouvêa, Renata Moro Sichieri, Maria Elena Bizelli, Sueli Javaroni, Maurício Rosa, Silvana Claudia Santos, and Geraldo Lima Sobrinho.

Cristina Esteley, from University of Villa María (Argentina) and Nilda Etcheverry, Norma Evangelista, Marisa Reid and Estela Torroba, from University of La Pampa (Argentina).

Anne Kepple, once more, for her insightful comments and editing of our "foreign English".

Marcelo C. Borba and Mónica E. Villarreal
August 2004

Chapter 1

WHY ANOTHER BOOK ABOUT TECHNOLOGY AND MATHEMATICS EDUCATION?

1.   INTRODUCTION

Computers have been a theme of intense discussion within the mathematics education community for more than two decades. If the notion of computers is extended to include other devices, such as calculators, it can be said that the debate has been going on for over thirty years. This being the case, a reasonable question might be: Why write about computers? This book represents an attempt to respond to this interrogation from various perspectives. One response could be that technology has not been used intensively in education, despite the efforts of a substantial part of the mathematics education community and the presence of an ever-increasing number of studies about computers, calculators, graphing calculators, and mathematics. Therefore, this book could be seen as an attempt to explain such a discrepancy.

Another possible answer could be that, over the last 35 years, technology has changed so much that constant updating is necessary. In Brazil, some researchers in the field of technology and mathematics education argue that the expression 'new technology' (*novas tecnologias*) should no longer be used, as computers and calculators have been around long enough to not be considered new anymore. On many occasions, we have twisted this issue around and argued that, since technology is changing so fast, we can always use the adjective 'new', especially if we concentrate on the notion of interface, seen as a means of relating to information and communication technology. Interfaces such as the video monitor, the keyboard and the mouse, which became popular in the 80's, are examples of how computer technology is renewed and transforms existing technology. Similarly, we can point to how flash technology and sensors, such as CBR[1] and CBL[2], began changing graphing calculators in the 90's; how newer versions of operational systems, such as Windows, became almost as friendly as the systems used for Macintosh since the 80's, resulting in the end of a 'direct contact' with the old-fashioned DOS system; how the Internet has changed computers, and how Java and www have changed the possibilities of the Internet. All of these changes have substantially transformed the way information technology can be used in education – transformations that we begin to examine in this book. Thus, this book could be seen as an attempt to examine how interfaces have changed the possible ways for one to learn in recent years.

Many authors, including us, have argued about the importance of coordinating multiple representations, now made possible with software that makes written, algebraic, tabular and graphical representations available. As technology has changed, and body movements can also be coordinated with standard representations in a more direct way, it has become possible to expand the notion of multiple representations. It can even be argued that the need exists to coordinate representations of the same type, for example, graphs produced by different media, such as computers and paper-and-pencil on the one hand, and body awareness of motion on the other. These different types of coordination have become part of knowing, and in this sense, it could be said that this book is about epistemological issues that arise when different technologies are associated with human beings.

Within the mathematics education community, one of the few issues on which there is consensus regarding the discussion about technology is that computers alone are not likely to bring any change, and that intense pedagogical discussion should be undertaken. In other words, if the decision is made to use technology in the classroom, the debate is still open regarding how to use them, from the perspective of the teacher and the students, as well as from the standpoint of other actors in the mathematics education landscape. We propose some pedagogical approaches in this book that we believe to be more resonant with these new technologies than others. In this regard, this book could also be seen as being about different pedagogies that could be used once it is decided that technology is relevant for education.

In spite of the 'free will' of teachers and administrators, whether they do or do not want to use technology, it has been hard to avoid using it due to

---

[1] CBR - Calculator Based Ranger - is a sonic movement detector that measures distances, velocity and acceleration.

[2] CBL is an interface that makes it possible for data, such as light intensity, temperature and electric tension, to be stored and transferred to the graph and table facilities of a graphing calculator.

social pressure from 'actors' such as politicians and business and school administrators. Such forces may use arguments such as "we must use computers because the labor force needs to be prepared for jobs in the future". In using computers not out of a conviction that they can be beneficial, but because it has been mandated, teachers may attempt to use them as little as possible, and when they do, make the minimum changes necessary in the structure of the curriculum and practices embedded in it. New technology in such an approach can be thought of as something that should not alter the *status quo* in school, nor 'touch' the way 'knowledge is transmitted in school'. New technology can therefore be 'domesticated'. Computers may be used as if they were 'electronic books', and graphing calculators as just a way of drawing graphs quickly. In this sense, this book represents an attempt to counter this way of conceptualizing technology in educational settings.

As suggested in the last paragraph, there is still an open question regarding the reasons why one should use technology. Twenty years ago, there was a debate between those who were opposed to using technology and those who were in favor of it because they believed it would improve teaching and learning (of mathematics). Although in most places technology is used to some degree in education, it could be the case that old questions regarding the use of technology are still open to debate, that issues regarding why one should use technology have gone unanswered. This book could then be seen as an attempt to propose another answer to why technology should be used in education.

This book is written for the mathematics education community, which could be loosely defined as being the set of researchers, professors and teachers who are interested in reflecting on the teaching and learning of mathematics, socio-cultural and political aspects of mathematics in school and society, and philosophical issues regarding the role of mathematics in education. Research in mathematics education has been gradually gaining identity in the last forty years or so. Most of members of this community like to debate theories about how one thinks and how one learns and teaches. In this book, we do not present a new theory regarding how to think about computers and education. We do, however, propose ideas, expressed in the form of theoretical constructs, about how we can overcome the dichotomy between humans and technology that underlies many of the difficulties that we, as a community, have experienced in implementing the use of technology in schools in ways that are not domesticated.

In addition to developing a theoretical discussion about the relationship between humans and computers, and what we call reorganization of thinking, we present several examples from research developed inside and outside the classroom. Examples help to shed light on the theoretical discussion. Local experience often has the power of becoming convincingly acceptable for different contexts. That is why when examples are presented, many of us say, "now I know what you mean!" or "This makes sense", meaning that we related to the example. As examples shed light on the theory, many may find reason to disagree with the ideas presented in this book, and to therefore help keep the debate alive in our field. Examples may, in many instances, be the bridge between those who like theoretical discussion and those who do not. After presenting examples, we will return to the theoretical discussion presented and introduce some new features into the debate.

The examples are, for the most part, from research conducted by a group led by the first author of this book. This research group, called GPIMEM[3] (Grupo de Pesquisa em Informática, Outras Mídias e Educação Matemática[4]), based at the State University of São Paulo (UNESP), Rio Claro Campus, São Paulo, Brazil, is one of the few in the so-called Third World that has been developing long term research about the use of technology in mathematics education, a theme which is often, due to prejudice, restricted to the so-called First World countries. This research group is composed of almost 30 members who are professors, associated researchers, doctoral and masters students, technicians, and undergraduate students. The undergraduates, who are engaged in a genuinely Brazilian program that is called Scientific Initiation, develop research at their level, under the supervision of professors, and receive, in exchange, a small scholarship. Their experience becomes a powerful item in their Curriculum Vitae, if they decide to engage in research after graduation. In our research group, all members meet periodically in different subgroups to solve particular aspects of different projects, or to think about the overall goals of the group. This book could, therefore, be considered to be about the activities of this group.

In GPIMEM, we use various research procedures and views of knowledge that are integrated. We can therefore say that we develop different research methodologies for different kinds of research, although all of them would fit within the so-called qualitative research paradigm. We develop teaching experiments as a means of documenting closely the way students, teachers and even workers deal with technology as they learn mathematics. Besides being a source for epistemological debate, the analysis of the data that come from these experiments has enabled us, as a group, to develop curriculum as we listen closely to students. Such experiments are integrated with studies developed in the classroom and in other landscapes

---

[3] http://www.rc.unesp.br/igce/pgem/gpimem.html
[4] Technology, other media and Mathematics Education Research Group.

where other educational actors are present. We do develop extension work, which has also become a setting for research. A major struggle of GPIMEM has been to articulate basic research, both in the lab and in the classroom, with implementation and research that can be applied more rapidly. Doing this without being trapped by pressures from the market, and being able to think about social transformation, has been a challenge. This book could, then, be considered to be about the interface between research methodology and implementation of research results with social concerns in mind.

In the above paragraphs, we have shown different reasons for a book about technology. A combination of the above topics is what the book is about. We present a discussion regarding theoretical issues; discuss examples, for the most part, from our research developed in Brazil; and raise questions about the problems of articulating research and implementing results in educational systems. We present a theoretical discussion that may help to convince some that pedagogy and curriculum should be changed substantially when qualitatively different media, such as information and computer technologies, are introduced in education.

In Chapter 2 we introduce the notion of reorganization of thinking and the idea that knowledge is always produced by collectives of humans-with-media. We present the notion that human thinking is reorganized by different media, such as computers and their evolving interfaces. By reorganization we mean that computers do not substitute humans, nor are they juxtaposed to them. They interact and are actors in knowing. They form part of a collective that thinks, and are not simply tools which are neutral or have some peripheral role in the production of knowledge.

In Chapter 3 we introduce our perspective of modeling as a pedagogical approach that has synergy with the use of information and communication technology. We do so contrasting it with the literature about problem posing and problem solving. The roots of modeling in Brazil are described and compared with the Danish project work. Different perspectives relating technology and modeling are presented. Finally, the limits of modeling are pointed out.

In Chapter 4, we present the experimental-with-technology approach. We discuss the meaning of experimentation in mathematics and mathematics education. We propose that modeling be used in conjunction with the experimental approach, an environment in which students raise conjectures, argue and 'prove'. The teacher has the role of coordinating the experiences of the students with what is traditionally accepted in academia. We also introduce a discussion about multiple representations and their relationship with media.

Within computer technology, visualization has taken on an important role, and this is why we devote Chapter 5 to this theme. We present an indepth discussion of the role of visualization in mathematics and mathematics education, stressing the value of this process in educational settings. We show how our perspective of humans-with-media as the basic unit that produces knowledge gives a new twist to the long time discussion about visualization in our community.

In Chapter 6, we present several examples from our research about modeling, showing how this pedagogical approach influences the use of technology by students, and how different technological actors, such as a function software or the Internet, and even paper and pencil, play a major role in the investigations developed by students in formal school settings. Almost all the examples come from classroom experiences in a mathematics course for biology majors where modeling was implemented as the pedagogical approach.

In Chapter 7, we return to the discussion of the experimental-with-technology approach and the process of visualization in educational settings, grounding it in examples from research. We suggest that experimentation and visualization are major attributes to be explored when computer technology is used. Examples in this chapter, as well as in the one preceding it, involve contents such as functions, derivative, integrals and associated topics from high school and early university-level mathematics curricula.

Chapter 8 is dedicated entirely to discussion about the Internet. Epistemological issues regarding the transformation of the notions of space and time that come with the presence of the Internet are addressed; as well as the nature of interaction in on-line education settings. Continuing education for mathematics teachers is emphasized in this chapter, as data from courses in which they participated are presented. Social issues, which permeate some of the examples of modeling, gain an important place in this chapter as we discuss how courses like this can be a path for giving access for teachers all over the country to a recognized mathematics education center in Brazil, like UNESP, the State University of São Paulo.

In Chapter 9, we apply to our own research group the notion of collective intelligence as we discuss how we integrate research, and the impossibility of an hierarchy in which the professor knows more, doctoral students know a little bit less, master students much less, and so on. We present a map of the research we develop, including the studies that are not discussed in detail in this book. Research methodology, procedures and epistemology are intertwined in this chapter.

Finally, in Chapter 10, we discuss the political dimensions of our research, as well as philosophical issues regarding the tension between psychological time, the pace of production of new technology, and elaboration of research about technology and mathematics education. We use elements from the history of education, regarding the introduction of the notebook in schools, to discuss the political role of having access to information and computer technology. We locate the work we develop in Brazil in the international scene, discussing our concern about democracy, the right of access to information and communication technology, and the importance of collaborative work with researchers from different parts of the world.

Chapter 2

INFORMATION TECHNOLOGY, REORGANIZATION OF THINKING AND HUMANS-WITH-MEDIA

1. THE SECOND INDUSTRIAL REVOLUTION AND EDUCATION

Almost twenty years ago, Schaff (1990), who originally published his book in 1985, discussed how the techno-scientific revolution would be changing our society. The author argues that his consideration would be valid only for the so-called industrialized world, and that analyzing the changes in "Third World countries" would be an even more complex task. Almost twenty years after he published his book, one may note that some of his predictions for the industrialized world are happening in less developed countries, as well. The "second industrial revolution", as Schaff calls the intensification of the use of computer technology, would provoke many changes in society. Among the changes that he predicted for the nineties and the beginning of this century, we want to highlight one associated with unemployment. Schaff foresaw that the loss of jobs provoked by the introduction of computers in many sectors of social activity could also bring about changes in the educational sector. Since knowledge is increasingly becoming the basis for jobs, and there will be a shortage of manual work, he proposed emphasizing continuing education as a means of solving the unemployment crisis. Although predictions about 'total' unemployment caused by the introduction of computers in different sectors of society have not completely come true, since information and communication technology have also lead to the creation of new jobs, Schaff's prediction regarding an expanding role for continuing education has proven true, since technology itself has opened up both possibilities and the need for life-long education.

Schaff (1990) is aware that this project is only possible if we can achieve a more democratic world, from an economic perspective, since this continuing education has a high cost. In such a perspective, work and education would become intertwined in one's life. Of course such a political perspective is not given. Alternatively, we would point out, and we think that this would agree with Schaff's premise, that an authoritarian perspective could possibly substitute this educational perspective of the near future, and that a 'knowledge-technology apartheid' could take hold. Some would have access to technology and others would not.

Continuing education on a large scale, and technology apartheid, seem to be at different ends of the political spectrum. Either we will be able to provide continuing education for the use of technology for all, or a divide between the ones who know how to deal with information and communication technology, and the ones who do not, will exacerbate the social crisis we have in the world. In order for us to help build the continuing education perspective for society, we need to battle on many fronts. For instance, in the political arena, we must struggle for this perspective and convince different sectors of various societies of its advantages, and try to mitigate the resistance of those who may need to forfeit certain economic privileges in order to fund such a perspective. As pointed out by Machado (1997), if this continuing education perspective becomes prevalent, the challenges for educators will be even greater than today, as more time will be dedicated to education. From within formal education institutions, we will have to emphasize pedagogies that stimulate the autonomy of learners so that, among other reasons, they can work in a world that is changing at an ever-increasing pace. This second concern will be addressed in Chapters 3 and 4, where we will discuss modeling and experimental perspectives for (mathematics) education.

A third dimension of such a struggle occurs at the epistemological level, which is the task we will tackle in this chapter. In order for us to implement Schaff's program, we need to understand the changes brought about in people's thinking when they are engaged in learning activities in which computers are available. If such a task is not taken up, we may fail to gain a deeper understanding of the kinds of educational changes that will occur if the educational-democratic alternative advances as computer technology becomes increasingly present in our lives. This kind of issue was not Schaff's concern, perhaps because he underestimated the transformations that this new media could bring about and failed to foresee that different people would also have different types of resistance to working with computers. In this chapter, we hope to contribute to this discussion, from the perspective of mathematics education, as we outline a framework of how we believe computers are related to people's thinking. It is important to emphasize that Schaff dealt with education in a more generic sense when he proposed a transition from a conception of *homo sapiens* to *homo studiosos*.

2.   REORGANIZATION OF THINKING

We begin this epistemological discussion by analyzing a paper by Tikhomirov (1981), a Russian psychologist, in which he discusses how computers affect human cognition and, consequently, how computers can change education. Although the original version of the paper dates from the 70's, his argument is relevant today. Tikhomirov argues that seeing computers as a substitute for humans is a shortsighted view. He claims that, although the output offered by a computer can sometimes be the same as that offered by humans, this does not mean that the program of the computer can be placed on the same level as human thinking, since the problem-solving skills are different. As a result of his research, he claims that:

> ... a large part of the control mechanisms of search in humans in general are not represented in existing heuristics programs for computers. When computer heuristics do resemble human ones, they are significantly simpler and are not comparable in any essential way. (p. 259)

He points out that one of these essential ways is that the process for directing a search for human beings is very different from the one used by computers. An issue that is not addressed by Tikhomirov is that his argument that computers will never substitute humans could be understood as being associated with the popular debate about computers taking jobs away from workers. It should be noticed, however, that the fact that machines replaced farm workers, and computers and robotics are replacing workers at various levels, does not mean that this kind of 'substitution' is equivalent to the substitution that Tikhomirov is referring to.

In discarding this notion of computers as a substitute for humans, Tikhomirov sets the stage to criticize another notion: that the computer is simply a supplement to humans. He discounts this notion by criticizing the information process theory of thought, which he claims is based on the premise that "complex processes of thought consist of elementary processes of symbol manipulation" (p. 260). If such an approach to thinking is adopted, the computer can be seen as simply increasing, or supplementing, the amount of information processed by humans - a view which presupposes only a quantitative, and not a qualitative, view of how computers influence human activity.

Moreover, reducing thinking to this view hides the fact that thinking also includes the goal one has in mind, the choice of the problem to be tackled, and changes in the problem during a process of investigation. Moreover, if one takes the information process theory as a paradigm, one fails to consider that context not only frames an activity, but also structures the activity, as suggested by Lave (1988). Needless to say, in such a view, values and politics are left out and understood as having no influence on thinking.

Conversely, if one agrees that values, politics and contexts 'shape' human thinking, it would be impossible to conceive of computers as mere supplements to mental activity, in the sense discussed above, as this would imply that thinking can be broken into pieces.

Before we analyze Tikhomirov's alternative to the theories of substitution and supplementation, one should consider why such a discussion is important. After all, information process theory is not that popular in mathematics education, and nobody talks about substitution and supplementation in our research community, either.

One answer to this question is that, although there has been much discussion about computers and cognition, and about the mediating role of computers in learning, there may still be some who believe in substitution or supplementation theory. In other words, the fact that human beings and machines are commonly seen as 'disjoint sets', and despite recognition that computers mediate the construction of knowledge, the 'cognitive unit' continues to be seen as just the human being, and not humans-computers, humans-paper-and-pencil, humans-computer-paper-and-pencil-orality, etc. This perspective can lead one to think of the role of computers as only supplementing humans, or merely juxtaposing humans, or even substituting them. The very idea of considering the human being as the unit that produces knowledge can underestimate the importance of technologies in this knowledge production.

Moreover, statements such as "computers develop students' thinking" and "computers help students to graph" may express a disjunction between humans and tools, depending on the theoretical framework used. We challenge the notion that paper and pencil should be considered as mere extensions of humans, and we claim that orality should also be seen as a technology and a medium. In Chapters 6, 7 and 8 we will return to this issue, basing our arguments on examples, and explaining the relevance of our interpretation of Tikhomirov's analysis for the mathematics education community. After this critique of these views of the relation between computers and cognition, we believe we have set the stage to analyze the alternative presented by Tikhomirov.

The alternative he presents to the theories of substitution and supplementation is that computers reorganize the ways humans know. Based on a Vygotskian perspective, Tikhomirov (1981) claims that the computer plays a mediating role similar to the one played by language in Vygotskian theory:

> We shall compare the process of regulating human activity through normal verbal commands with the process when aided by a computer. The similarity here is obvious, but there is an important difference: possibilities for immediate feedback are much greater in the second case. In addition, the computer can appraise and provide information about intermediate results of human activity that would not be perceived by an outside observer ... Thus, with regard to the problem of regulation we can say that not only is the computer a new means of mediation of human activity but the very reorganization of this activity is different from that found under conditions in which the means described by Vygostsky are used. (p. 272-273)

The above quotation suggests that Tikhomirov bases his theory of reorganization on Vygotsky's notion of regulation by language and on the argument that regulation provided by computer technology is qualitatively different when compared to that provided by language.

If one considers that computers in Tikhomirov's time had interfaces that were less user-friendly than the ones in use today, and that feedback was not as fast, one can think that the change in quality he discussed was undergoing quantitative transformations, as well, and that maybe new qualitative changes have already taken place, considering the type of feedback that has become possible as robotics, sounds and other kinds of interfaces have transformed computers.

Visualization has been the main change in computer interfaces since monitors were introduced as an essential part of computers. These changes have increased accessibility to computers, making them available to a larger audience, and also affected this feedback, in a way Tikhomirov could not have predicted at the time. Later in this book, the idea of reorganization will gain new dimensions as we present and discuss examples.

Tikhomirov does argue that there are problems in which the computer 'can' substitute humans. A problem that has been written in the form of an algorithm, and incorporated into computer software, can mimic the solution developed by humans. However, although the results may be the same, the processes are different. He wants to emphasize that the computer's solution has incorporated creative work in the form of an algorithm. Moreover, for users who are going to have access to such a program involving this algorithm, the experience will be very different from the programmer's. In other words, the use of the program occurs through the interface, which changes the nature of the experience of the user and of the programmer who knows the algorithm for a given program. So even such an activity, which can be seen as substitution, from the point of view of the user, is a new human activity. It is an activity that has been reorganized. Machines do substitute humans as, for example, in a car factory, but they are embedded with human algorithms, and they demand a reorganization of thinking by humans.

This is precisely what is important for education. Knowers deal with software developed by professionals of varying origins, and they have to deal with this stored information, with this new type of memory. If computers reorganize the way humans know, could it be inferred that the introduction of computers in education itself will provoke change? In other words, can it be concluded from Tikhomirov's discussion that computers 'per se' lead to change in education? If the answer is "yes", how can we relate such a conclusion to a near consensus within the mathematics education community that computers by themselves are not enough to change educational practices? If the answer is "no", how should one make sense of Tikhomirov's idea?

One way to answer the above questions is to differentiate thinking from education. A computer can provide the same solution to a problem that a human would. This, as the reader may be convinced by now, is not the same as solving the problem in the same way. The focus of our attention should thus be directed to what happens when humans-computer systems solve problems. In this sense, when features of the computer are incorporated into the search process of humans, when the 'search' process of the computer is used in combination with that developed by humans, and the various forms of feedback provided by the computer contribute to the emergence of new problems, one can apply the metaphor of reorganization to describe what happens to thinking. If computers are used to solve problems and/or generate new problems, one can say that reorganization of thinking has occurred, but of course this reorganization does not preclude a pedagogical discussion. The pedagogical argument, as we will see later on in Chapter 3, shows how the different forms of reorganization of thinking may occur in educational processes. Thus, what has been discussed so far is not at odds with one of the few issues on which there is a consensus within the mathematics education community. On the other hand, this discussion tries to emphasize what may not have been highlighted enough in this community: the idea that changes in educational practices should take into account this reorganization of thinking and the solution of problems by humans-computer systems.

3.   INTERSHAPING RELATIONSHIP: STRESSING BALANCE

Vygotskyian theory has emphasized language as a means of mediating human activity, as discussed extensively in the literature. Tikhomirov has taken this idea further as he conceived of computers as being a qualitatively new kind of mediator. The emphasis, however, seems to be on the influence of the computer on the knower. Authors such as Noss and Hoyles (1996) also emphasize the role of computers, or any tool, as mediators of knowledge:

> Focusing on technology draws attention to epistemology: for new technologies - *all* technologies - inevitably alter how knowledge is constructed and what it means to any individual. This is true for the computer as it is for the pencil, but the newness of the computer forces our recognition of the fact. There is no such thing as unmediated description: knowledge acquired through new tools is new knowledge, MicroworldMathematics is new mathematics. (p. 106)

This notion, which is one we share, stresses the influence of tools on the way one knows and what one knows. Borba (1993), for example, comparing the manipulation of graphs with paper-and-pencil and with computers, claimed that:

> Consistent with the medium used, the main task of traditional curricula was to sketch the changes caused in a graph due to changes in the algebraic expression of a function... The medium, then, has 'shaped' the mathematics and the thinking of those who have mathematical training, despite the fact that paper and pencil were always taken for granted as 'neutral' tools. (p. 3)

This notion has the importance of making clear the social influence on the way one knows. However, it does not give enough emphasis to how tools are sometimes used in ways other than those they were designed for. As a counter-balance to this notion that tools influence the knower, Borba (Borba, 1993, 1995a, Borba and Confrey, 1996) has proposed the notion of the intershaping relationship, intended to be a more symmetrical notion to describe the relationship between knowing and technology. This notion evolved in response to a felt need to express something his students (the subjects of Borba's research) were teaching him, and to find an expression that could fill the gap between the notion of mediation on the one hand, and the notion of 'construction of knowledge by a knower' on the other hand. Some examples are presented in Borba (1993) of students interacting with the technology in ways the designers could not have predicted. The notion of intershaping relationship intends to express a balance between the shaping of the knower by the socially historical medium available and the shaping of this medium by the knower. "The cycle between representations and students' construction could be endless in this 'intershaping relationship' " (Borba, 1993, p.8), although it is noted that "this 'intershaping' relationship is not predictable and mechanistic" (p. 333). Therefore, any suggestion that human thinking could become standardized by a given media would not be defendable, as different and unpredicted uses of a given medium could always take place. Media, therefore, condition the way one may think, but do not determine the way one thinks.

When Borba was developing the notion of the intershaping relationship in the early 90's, he was influenced by the heated debate in the education community in the U.S.A. regarding Piaget and Vygotsky, and whether their ideas could be merged or were 'incommensurable', and therefore impossible to integrate. He was also heavily influenced by his doctoral research and activities in the mathematics education research group at Cornell University. At that time, this group, led by Jere Confrey, with the strong influence of Erick Smith, had spent years designing and developing a software called Function Probe[5]. In the long design meetings, mathematical ideas and the way they should be incorporated were discussed exhaustively. Many attempts to model possible reactions of the students (future users) to the software were made through teaching experiments (Cobb and Steffe, 1983), through extensive review of the literature of the subject matter, functions in this case, and through the direct input of students who used early versions of this software in university-level pre-calculus courses. Although we felt that we knew a considerable amount about the software and students' interactions with it, we were constantly surprised by how students were always able to find new ways of dealing with it, as long as the commitment to letting the students' voices emerge remained (Confrey, 1994). In Borba (1993, 1995a), in particular, we are able to detect a series of examples in which students used the software in ways that were completely foreign to the concerns of the design team. In other words, students had shaped the software, and had not only been shaped by this tool.

Over the last several years, as we have re-examined this notion, we have brought ideas related to dialogue, hermeneutics, and technology to further

---

[5] Function Probe (Confrey, 1991b), is a multiple representation software with capability to link tables, graphs and algebraic expressions. It allows for the generation of graphs through the mouse and icons on the monitor, bypassing algebra as a mandatory step in a given path. Function Probe was developed to be a software that would make it possible to teach function starting from visualization, with the capability to manipulate graphs "directly" through the mouse, and with links to table and algebra windows that opened doors to establish relationships between graphs, tables, and algebra.

explain what Borba wanted to stress with the notion of intershaping relationship. Garnica (1992) has emphasized the dialogical relationship between the student and the teacher, and also between the reader, whether a student or a teacher, and the text. He proposes a pedagogical approach that brings hermeneutics into the classroom, in which the meanings of the written words in a mathematical text are explored. Garnica's work, together with Borba's earlier research (1987, 1997a), inspired by Freire (1976, 1981) and Schutz (see Wagner, 1979), brought the idea of using the notion of a dialogical relationship to analyze the relationship between a user and a piece of software at a micro level. Although we consider the humans-computer system to be the basic unit of analysis, we acknowledge the importance of examining the relationship within this unit, and propose that there is a dialogical relationship between a software user and the intentions of the group or person who designed the software. It is clearly a different kind of dialogue than the one established between two persons, in the same way that a dialogue with a text is also different. But the intention of the software designers, their desire reflected in the interface, their pedagogical goals, and the intentions and desires of the user permeate this dialogical relationship. As a result of this 'intra-relationship', we can claim that the use of computers will lead to an even greater diversity of ideas compared to when humans did not had access to this medium as part of a basic cognitive unit.

If such a view is taken, technology can be seen as having a strong human component since software, a technical product, has a strong human influence on its design and on the interfaces between the software and the user. Therefore, ideas such as Lévy's (1993), that we should not create a dichotomy between humans and technology, should be analyzed more carefully. Technology, in general, as we will discuss in the following sections, is seen as having a human dimension (the design, the choice of knowledge to be stored, and so on), and humans are seen as having a strong technical component as well.

4. HUMANS AND TECHNOLOGY: A HISTORY OF SEPARATION

In looking up the terms technics, technique, technical and technology in some philosophy dictionaries (Lalande, 1996; Japiassú and Marcondes, 1996; Brugger, 1969) one may notice some commonalities. The invariant in those definitions is the opposition between technics, or technology, and humans. Humans are users of technology, they build or develop technology, but humans and technology are always seen as disjoint sets.

Let's analyze some of the definitions presented in these dictionaries to illustrate how this division between humans and technology may be inappropriate for understanding the role of computers in society, and in education, in particular. A more generic definition of technics can be found in Lalande (1996), who defines it as:

> a set of well-defined and transmittable procedures aimed at producing certain useful results. Another noteworthy characteristic of this initial technics, which was the infrastructure on which physical science rests, is its permanence over the centuries.... It is an institution, probably the oldest of the institutions, and still remains, with the same characteristics today that it had at its beginnings... they are traditions that are passed from generation to generation by way of individual teaching, learning, and oral transmission. (p.1109)

Although this definition can be seen as somewhat outdated, in view of how transmission among humans is described, it should be noticed that the role of humans becomes clearer only when the idea of education is brought into the discussion. In other words, humans generated these technics, but this is not emphasized in the above definition, and only the role of cultural transmission, one aspect of education, is attributed to human actors. Prior to that, technics is described only as a set of procedures. Later in the same item, Lalande proposes the idea that technics is the "set of individual procedures of an artist, or writer" (p. 1110). Although this definition was the one that associated humans most closely with technics, in that it links the term with arts and writing, which are usually accepted as very human activities, it should be noted that technics is seen as something mechanical, procedural, uncreative. In other words, technics is seen as something mechanical, as opposed to humans, who use technics (such as the technique of the gothic style) as the basis for developing their creativity. The same emphasis on the mechanical aspect of technics can be found in Japiassú and Marcondes (1996) who define it as a "set of rules or procedures adopted in a job aiming for predicted results. Practical ability." (p. 257).

In the same dictionary (Japiassú and Marcondes, 1996), it is emphasized that, in the classical Greek approach, there was no interaction between science and technics, and that in a modern sense, technics is seen as a product, as a practical application of science. The same point is found in Lalande (1996), in the definition of 'technical', where he attributes this notion to Cournout[6], and points out that only Kant places the theoretical and

---

[6] Cournot, Antoine Agustin (1801-1877), was a French mathematician, economist, and philosopher. In *Recherches sur les principes mathématiques de la théorie des richesses* (1838), he introduced the application of mathematics to the study of economic problems.

the technical in close approximation to each other. Also found in this last reference is a definition of technology as being "the study of technical procedures" (p. 1111) or as theory about technics. It is also pointed out that technology is often used to replace what has been discussed as being technics. In the dictionary by Brugger (1969), in the discussion of the word 'technical', the opposition of human beings versus technical products is strongly emphasized. He raises the issue of the domination of humans by machines. It is argued that technics can 'serve' humans or can dominate humans. Although this entry emphasizes the technology-humans dichotomy, it is also the one that gives most attention to the relation between technics and humans, arguing that the former can come to dominate the latter. If this happens, it is argued that humanity could become enslaved by machines.

We did not intend, with this short excursion into a few philosophy dictionaries, to develop a summary of philosophies of technology. We intended only to illustrate how the dissociation between humans, on the one hand, and technics, technical and technology, on the other, is deeply rooted in the philosophical perspectives expressed in many dictionaries; how technics has been viewed as mechanical while humans are creative; and how technics has been described as oppressive and humans as liberating. For the purposes of this book, we will build on the definition of technology proposed by Kenski (2003): "the set of knowledge and scientific principles that are applied to the planning, construction, and utilization of a piece of equipment in a specific type of activity" (p. 18). Since we adopt a notion that knowledge is always human-bound, we believe that the dichotomy is overcome. Later in this chapter, we will associate humans and technologies even more closely.

Although the authors of the above references, with the exception of Kenski (2003), fail to mention information technology, it may be the case that this traditional way of thinking about technology may influence the fears of many educational actors regarding its use in pedagogical processes. Our argument is that part of the silent, as well as explicit, resistance to the use of technology in schools may be partially rooted in a philosophical view that dichotomizes technology and humans, and privileges the latter over the former, except when technology is not perceived as a threat to the *status quo*, and people feel assured that this 'new' technology will not substantially change the school hierarchy or the content taught in class. We want to raise the hypothesis, which requires closer examination, that part of the fear of incorporating technology in the classroom is based on the dichotomy

---

He was the first to formulate a complete theory of the monopoly. (Japiassú and Marcondes, 1996).

discussed throughout this chapter. Such a dichotomy would provoke fears in those who consider themselves to be separate from technology, and who do not see themselves as being in a process of constant change, shaped, in many ways, by technology.

But the obvious question then becomes: What is the alternative? How can we view technology in a way in which it is not so dichotomized? In the previous discussion in this chapter regarding Tikhomirov (1981) and Borba (1993), a slight distinction between computers (a type of technology) and humans could be perceived in the views of both authors. Noss and Hoyles (1996) described the mediating role of the computer in mathematical learning as structuring, and being structured by, the activity of the learners, when a particular type of software is used. They assert that: "The computer provides a screen on which learners can express their thinking, and simultaneously offers us the chance to glimpse the traces of their thought" (p. 6). The authors emphasize that the computer is not the focus; rather, they are interested in "what the computer makes possible for mathematical meaning-making" (p. 5). That is why they propose seeing the computer as a window on knowledge, conceptions, and attitudes of students and teachers. Therefore, it seems to us that the computer is still seen as separate from humans. Borba developed the notion of the intershaping relationship to emphasize the way computers shape knowers, and how knowers shape computers. It was a way of balancing the influence of computers on the knower with the influence of the knower on technology, and also a means of avoiding any type of a return to neo-behaviorism and the idea of the tabula rasa. This notion can also be seen as a way of trying to overcome the objective-subjective dichotomy[7], and moving beyond the idea that cognition is something solely 'internal' to human beings. It seems that the germ of such an intention was even more evident in Lave's (1988) work. When she takes as her basic unit of inquiry 'the person-acting (in setting)', she definitely moves a step forward in this direction, however the separation remains between the person knowing-in-the-world and the world or setting. Tikhomirov, Noss and Hoyles, Lave, and the former work of Borba, dispense with the dichotomy but maintain a separation, in the sense that there is an underlying implication that only humans produce knowledge. In the next section, we look at an alternative proposed by Borba (1999b, 2002), based on the work of Lévy, to overcome the humans-technology dichotomy discussed in this chapter, and argue how it can be useful for mathematics education. Moreover, we think that Lévy's work can be a 'scaffolding' for the efforts to overcome the dichotomy mentioned above - an effort which some of us in the mathematics education community are engaged in. We

---

[7] See Ernest (1991) for a discussion of objective and subjective knowledge.

believe that this way of understanding the relationship between human beings, cognition, knowledge and technology may shed some new light on some topics in mathematics education.

5. MEDIA, HUMANS AND KNOWLEDGE: POSSIBILITIES OF MERGING

The bridge proposed by Lévy (1993), of introducing an alternative view of cognition that can account for the power of this new media, the computer, represents a critique of the way many authors have dealt with the notion of technics, emphasizing the dichotomy between humans and technology. In creating such a separation, political dimensions of technology may become obscured, because democracy, commonly seen as a human endeavor, and technology may be seen as disjoint sets. Moreover, as we will try to argue, such a separation can lead to the view that knowledge is produced solely by humans, and can be translated into another medium without being affected. Such a view can lead to the conclusion that few changes will be necessary for curricula as computers - and their constantly updated interfaces - become the 'hegemonic medium' in learning, an idea we disagree with, as will become clearer later on.

Technology is still perceived in some educational sectors, in different parts of the world, as a threat to humanity. In this view, computers may de-humanize humans, may take over society, and may not let students learn what they should. Even some who accept the use of computers in education, for instance, may argue that they can be used as long as they do not change anything in what was learned before. In other words, there seems to be an ideology, which contrasts human beings to computers, that claims 'to protect humans from the attack of machines'. But as Lévy (1993), argues, the effect of such ideas may very well be the opposite: humans may be disempowered by such a notion, since they feel impotent in the face of technology, and may think that there is no chance of achieving any kind of democracy within this new 'technological order'.

Lévy (1993) argues that people who claim that new technologies are dangerous to humans are not aware that the medium they are using to express such ideas - either orality or writing - is also a medium which structures their practice[8]. In other words, writing has become invisible for

---

[8] Lévy uses a notion of structuring in a way that is very similar to Lave (1988). In analyzing both authors, it seems that the common root of their work is Bruno Latour's writings. Latour's work (1987) also seems to be the basis for the notion of cognitive ecology that is proposed by Lévy in many of his books. These three authors have a common reference to

social actors, and they cannot see the intershaping relationship between them and an 'inoffensive medium' such as orality or writing. As they are not able to account for that influence, they believe that cognition was developed independently of media.

Lévy (1993) emphasizes that media have always been intertwined with the history of humanity. He uses the notion of technologies of intelligence to characterize three main technics that are associated with memory and knowledge: orality, writing, and information technology. In this sense, orality was used to extend our memory. Myths were a way developed by societies to store important parts of their cultures. The diffusion of written texts that began in Europe at the end of the 15th Century with the appearance of the book, in a format similar to the one we have today, made it possible for memory to be extended in a qualitatively different way than it had been with orality, another technology of intelligence. Thus, writing emphasizes linear reasoning and allows thinking to be shown in a linear fashion. The logical sequences and the narratives, although already in existence before the popularization of writing, only gain importance with the change in technology that increases accessibility to books, paper, pens, and associated tools.

Information technology should be understood in the same way. It is a new extension of memory, with qualitative differences in relation to other technologies of intelligence, and makes it possible for linear reasoning to be challenged by other ways of thinking, based on simulation, experimentation, and a 'new language' that involves writing, orality, images, and instantaneous communication. In this context, the metaphor of linearity is increasingly substituted by the discontinuity that characterizes the use of the Internet. For example, every time someone accesses a given home-page using a 'link', or uses more traditional software dealing with geometry or functions, accessed through the menu, we are experiencing discontinuity. This challenge is also faced by those who are investigating this 'new language' that results from the almost daily appearance of new interfaces in information technology media.

This brief summary of the history of media makes it apparent that the dichotomy between technology and humans is not based on a historical perspective like the one presented above. The perspective we embrace suggests that humans are constituted by technologies that transform and modify their reasoning and, at the same time, these humans are constantly transforming these technologies. From this perspective, a dichotomous view

---

the work of the English anthropologist Jack Goody, especially his book *The domestication of the savage mind* (Goody, 1977).

does not make sense. Moreover, we believe that knowledge is produced together with a given medium or technology of intelligence. It is for this reason that we adopt a theoretical perspective that supports the notion that knowledge is produced by a collective composed of humans-with-media, or humans-with-technologies, and not, as other theories suggest, by individual humans alone, or collectives composed only of humans.

In this sense, we believe that humans-with-media, human-media or humans-with-technologies, are metaphors that can lead to insights regarding how the production of knowledge itself takes place, in the same way that the human being is also a metaphor for the epistemological subject that is so deeply rooted that it is assumed, by many, to be natural. But why a new metaphor? Will it become just another name? We believe that the answers to these questions lie in the reader. In our opinion, this metaphor synthesizes a view of cognition and of the history of technology that makes it possible to analyze the participation of new information technology 'actors' in these thinking collectives[9] in a way that we do not judge whether there is 'improvement' or not, but rather identify transformations in practice. In other words, this notion is appropriate for showing how thinking is reorganized with the presence of information technologies, and what types of problems are generated by collectives that include humans and media such as paper-and-pencil, or various information technologies. Lévy also proposes that other kinds of technologies, such as cities, and libraries, are also actors in the production of knowledge. In this book, however, we will focus on changes provoked by technologies of intelligence. In particular, we use medium and technology almost synonymously, as we believe that technology is always used to communicate, and that a medium can always be seen as a technology. We often use the term medium to emphasize the communication aspects of technologies of intelligence.

Media are undergoing constant change, as increasingly user-friendly interfaces are created within collectives of humans-with-media. To illustrate this assertion, we return to Lévy (1993) once more. He presents an example related to the history of books. The predecessor to the book was the scroll, which, in addition to its inconvenient format, was also full of notes and comments that made it difficult to manipulate. As books, which first appeared in the 15th century, began assuming the form they have today, the path was opened up for their popularization – a path that was completed when characters were invented to occupy less space in books, decreasing the cost. According to Lévy, printing thus became a springboard for the 'explosion' of science that took place later on.

---

[9] *Thinking collective* is a term used by Lévy to emphasize that knowledge is produced by collectives composed of human and non-human actors.

A similar point can be made regarding the history of computers: interfaces and cost were key factors in the role that computer technology plays today. Another comparison than can be useful for our discussion can be drawn between newspapers and computers. Different medium extend our memory in different ways, and reorganize it in distinct manners. When one has a text, or several texts, in a word processor, e-mail manager, or Internet browser, one has the freedom to navigate from one text to another, make changes and so on. On the other hand, the user is constrained by the size of the screen, in that he/she can only see a small piece of the text, and is also unable to actually touch the text or manipulate it in the traditional sense. In contrast, one can touch and hold a newspaper, and can get an overview from skimming the first page, or the headlines of all the other pages.

What is important to note in both examples is that the change of interfaces of books or of the computer medium changes the way we have access to information provided by others, and can decide whether more or less people will have access to such information. Moreover, technical development generated by humans is the key to such changes, which may lead to a more human view of technology. In this, technology and humans constitute a unit, and if they are viewed separately, it can lead to problematic views with respect to education. For instance, many who favor the use of computers in education try to protect academic mathematical knowledge from any kind of change. In such a view, mathematical knowledge is independent of the medium, and the job of the educator is to try to use the computer in a way that does not affect mathematics. We believe that this leads to the domestication of this new medium, and efforts to reproduce practices and styles from other media. We claim that a new technology of intelligence results in a new collective that produces new knowledge, which is qualitatively different from the knowledge produced by other collectives.

Once writing was introduced, orality was also transformed. As memory was extended to paper, it was possible for theories to be born. In the case of mathematics, the opportunity emerged for long demonstrations to be developed and stored. It is relevant to note that writing did not abolish orality. On the contrary, it created what Lévy (1993) labeled secondary orality, which would be orality related to reading what has been written. In the same way, as will be argued later on in this book, the computer created new forms of writing and orality.

The connection between humans and technologies of intelligence is stronger than one might imagine, and Kerckhove (1997) proposes that the connection between body and technology is closer still. He sees the alphabet as a technology that shapes the development of our brain, and claims that users of the Roman alphabet read from left to right due to cultural factors and to the way our vision functions. For instance, he invites us to look at the two rectangles below and answer the question: Which of the dotted lines in the rectangles in Fig. 2-1 goes down and which goes up?

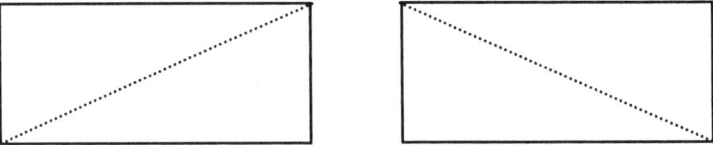

Figure 2-1. Rectangles and alphabet

Most of us who use the Roman alphabet, and read from left to right, will say that the line in the rectangle on the right is going down and the one on the left is going up. This author developed extensive research in neuroscience and psychology, proposing that those of us who are impregnated with the alphabet see the world in different ways. As we learn to sequence letters and sounds, we do this with all other things, as well, including the way we see space. For this author, there is nothing natural about the notion of perspective, and he asserted that it was invented to order space with the larger figures being seen first. He links neuroscience with the history of writing to claim that all the "writing systems that represent sound are written horizontally, but all those that represent images, such as the Chinese ideograms or the Egyptian hieroglyphs, are vertically written" (Kerckhove, 1997, p. 60). This author claims that the connection between mind, body and technology is much closer than had been believed. For the purpose of this book, it is important to consider that alphabets, speech, images and the new type of 'language' provided by computer interfaces are not strictly external to us; they are also internal, a part of us. Conversely, humans are present in and impregnate these technologies of intelligence.

We may use the idea of a socio-technical network, as proposed by Lévy (1993), as a metaphor to understand the relationship between humans and machines. This network can be seen as a hypertext in which changes developed by humans in the way they use a technology, and changes in technology, provoke changes in the way collectives of humans-with-media understand the world.

In such an approach, technology, medium and interface play a key role. An interface can change a medium in such a way that it generates a completely new medium. Take as an example, computers in the 70's with their cards, lack of monitors, large size and generation of heat. The way we interacted with those computers was very different from the way we eventually communicated with the Macintosh computer, which brought us the monitor, mouse, pull-down menus, and different icons, and later with

PC's. In other words, the change of interface dramatically altered the medium, the way humans related to it, and the nature of humans-computer systems as the unit that produces knowledge. These systems undergo dramatic changes when a new 'peripheral piece' is introduced. This was the case of the laser printer, the scanner and the automatic translators and the www interface.

If technology is understood in the way discussed here, it can be the basis not only for educational practice, but also for an epistemology that stresses the role of media. Knowing in this sense becomes, therefore, an endeavor not only of humans, but of a collective of humans and things. Particular attention is paid to humans and to the technologies of intelligence due to the more evident role they play in our ways of knowing.

Lévy (1998) builds on this notion to present a democratic perspective of knowledge that is called 'collective intelligence'. Thinking collectives and collective intelligence are connected. Our individual consciousness and cognitive processes are always subject to interaction with the technologies of intelligence. That is why we support the notion that humans-with-media (or humans-with-technology or humans-with-technology-of-intelligence) should be the basic unit of knowledge. This collective, formed of humans and non-humans, produces meaning as it connects different nodes of a network. Network of meanings is the metaphor for how this collective of humans-with-media produces knowledge. But we can also think of an intelligence that is collective. In such a view, one intelligence does not compete with another. They collaborate! Different combinations of humans with media, located in different parts of the world, gain power in some domain that becomes part of this collective intelligence. The ideas of the carpenter and of the mathematician complement each other. An intelligence like this is thought of as being a complex network. The Internet is a good model for it, and at the same time, it makes it possible for us to take advantage of collaboration on a larger scale.

In such a perspective, ignorance can even be praised, in the sense developed by Kerckhove (1997), who claims that being aware of what one does not know, and the possibility of knowing, searching and forming collectives, can transform humans into more flexible, empowered beings than those who claim to 'know' and have more programmed ways of thinking. Intelligence is collaboration, so we should always praise knowledge that is different from our own. Lévy and Kerckhove, who are from France and Canada respectively, seem to have much in common with the ethnomathematics movement that was born in Brazil. Neither sees scientific knowledge as the only way of knowing, and they both challenge the assertion that there is an absolute way of knowing. It is interesting that the perspective of ethnomathematics (D'Ambrosio, 2001), which has tended to focus on peoples whose ways of knowing have been suppressed (e.g. slum dwellers, landless peasants, indigenous peoples from different parts of the world) is similar to the perspective of these authors (Lévy, 1993, 1998; Kerckhove, 1997), who focus on recent technologies, like the Internet and interfaces such as www, or technologies yet to be invented.

As the reader will see, we will argue that perspectives in mathematics education, such as ethnomathematics and the one adopted in this book regarding the connection between technology and humans, have more in common than one may think at first. In this chapter, we have discussed some views regarding the relationship among humans and technology. Based on various authors, we have proposed that we consider humans-with-media as the basic unit for thinking. We believe this view can be the basis for an epistemology that focuses attention on how people know things in different ways with the introduction of different technologies. We believe that this view may help us see that knowledge has always been conditioned by different media throughout human history, but that for the first time, as Lévy (1993) claims, we have the chance to consciously interfere in the way this technology may shape our life. We should not think that technologies determined the way we know mathematics, but we hope that after this discussion, it will be possible to see that it would be unwise to think that mathematical knowledge, and the way we know, have not changed over the last few decades.

We have built on the views of Tikhomirov and Lévy to present the notions of reorganization of thinking and of humans-with-media. In Chapters 6, 7 and 8, we will discuss some examples from our research group to illustrate how such a view has been materialized. Before doing so, however, we will open up another perspective in the next chapter as we discuss how what we, in Brazil, call modeling, and the Danish call project-work, can be a rich perspective to be adopted in education, particularly when computer media have been transformed into, among other things, devices for conducting research, as well as subjects of research.

Chapter 3

MODELING AS A PEDAGOGICAL APPROACH: RESONANCE WITH NEW MEDIA

1.   INTRODUCTION

In the last chapter, a theoretical view regarding the role of computers in mathematics education was presented. However, the relationship between pedagogy and technology was not emphasized. Penteado Silva (1997) has illustrated how the arrival of 'new actors', such as computers, causes an impact in a school. As she focused on the role of teachers in such a process, she was able to document and interpret the struggles of teachers to either maintain their pedagogical practices unaltered, using paper-and-pencil, or to find new practices that would be suitable to the new medium.

In this chapter, we will discuss one of the pedagogical approaches that we believe is in resonance with computers. We will present a pedagogical approach - which has been named *modeling* in Brazil and *project work* in other countries, such as Denmark - which we believe becomes even more powerful with the use of new technology and can bring substantial change to curricula developed inside and outside the classroom. Modeling can be understood as a pedagogical approach that emphasizes students' choice of a problem to be investigated in the classroom. Students, therefore, play an active role in curriculum development instead of being just the recipients of tasks designed by others. Before presenting the discussion about modeling itself, we will introduce the theme in a framework that relates pedagogy to epistemology.

Lincoln and Guba (1985) use the notion of resonance to support the idea that there are research methodologies that are in harmony with different conceptions of knowledge. They argue that values are embedded both in different paradigms of knowledge and in methods, and if this is not taken into account, some research may be pointless. One of the examples they use to illustrate their point of view is the case of how to introduce reading to kids. The debate "whole language versus 'decoding' or 'skills' models", a discussion which was relevant in the 1980's and still is in many countries, is appropriate for the discussion being presented in this chapter. The decoding or skills models are based on the notion that little pieces of knowledge can be gradually assimilated by a learner. There is an underlying notion in such a view that thought can be fragmented. Reading is seen as a product that can be diagnosed as faulty. On the other hand, in the 'Whole Language' approach, reading is seeing as an extension of speaking, and the emphasis is on experience. The child's environment and the spoken language are the basis for this pedagogy.

The question raised and the answer given by Lincoln and Guba (1985), following their discussion of differing approaches to reading, are:

> Which paradigm is resonant with which view, and which is dissonant? If one is guided by the decoding or skills models ... it seems clear that the conventional inquiry model provides a good match. The very notion of "decoding" or skills fits well with a view of a tangible and fragmentable reality. Identifying and studying reading variables and their relationships seems quite appropriate. With structure and product as foci, research aimed at devising generalizations and cause-effect statements may not appear to be entirely aberrant. ... In short, the value underpinning of the conventional research model is very similar to the value underpinning of the decoding or skills model ...
>
> As we move to the whole language model, which views reading as a process ongoing in the learner's head in interaction with his or her environment and in view of earlier experience, we see that a research method that requires breaking phenomena down into variables and their relationship has little to recommend it. Generalization has little meaning when one is dealing with such idiosyncratic dimensions. Cause-effect relationships can hardly be sorted out when they are simultaneous and interactive. (p. 180-181)

They establish a very powerful relationship between research methods, epistemology and pedagogy. A pedagogical approach is embedded in a view of knowledge, and so are research procedures. If a given pedagogy is assessed by some procedure that is dissonant with its underlying view, very little can be established. Test results have little to say to theories in which process is more relevant than products, and qualitative data will have little value in views of knowledge in which a product is paramount in order to construct an ever-increasing accumulation of knowledge.

In mathematics education, it is easy to find examples in which the dissonance between pedagogy and assessment methods causes immense trouble. The most well-known case may be the "California Framework" in the U.S.A. Many states in this country, including California, followed the recommendation of the N.C.T.M (National Council of Teachers of Mathematics) and other associations to bring less rote learning, and more problem solving and open-ended activities into the mathematics classroom. Even textbooks were designed to encourage such an approach, and were adopted by some public schools in that country. On the other hand, tests were designed as they always had been, and used to assess such initiatives. Assessing the new pedagogical approach based on the California Framework, using the old methods of evaluation, generated incongruencies. Although the situation is obviously far more complex, involving electoral politics and, struggles between mathematicians and mathematics educators, it is reasonable to say that the dissonance between the underpinnings of the evaluation procedures and the pedagogy that was being implemented has contributed to a new wave of 'neo-back-to-basics movements', in which kids are expected to know the 'mathematical facts' by heart.

Meloy and Barros (2000) give a nice example where three geometry students in a Boston (U.S.A.) public school discussed what happens to the area of a rectangle whose base is fixed and whose sides remain the same length when it is transformed into a parallelogram by pushing the upper corner of the left-hand side towards the right. This discussion took place immediately after a traditional quiz where students were asked to calculate the area of a parallelogram like the one in Figure 3-1

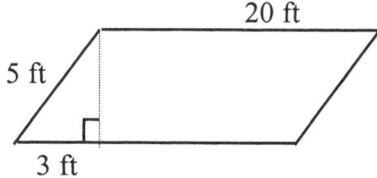

Figure 3-1. A parallelogram to calculate its area.

After the quiz, the teacher asked the students to form groups and help each other to understand their mistakes. One of the groups was composed of three students, two of whom did not complete the above problem and failed the test because, although they knew the formula to calculate the area of the parallelogram, they could not find its height. The third student was successful in this item. During the discussion, one of the students who could not answer the question showed a deep understanding of the meaning of the area of a parallelogram and the way it varies when a rectangle is transformed into a parallelogram with the same length and width as previously described; meanwhile, the other students asserted that the area of the parallelogram would be the same. During this group activity in class, the teacher (one of the authors of the article) quizzed and tried to understand the discrepancy between the student who could solve the problem with the formula but did not understand the variation of the area as a function of the height, and the other student, who understood but could not solve the problem during the test.

This example illustrates that knowing how and when to apply a formula does not necessarily imply understanding, and that although a rich discussion was held around the concept of area, the assessment was based heavily on products (correct answers to given questions), showing a kind of dissonance between the evaluation procedure and a pedagogy in which the key was not finding correct answers, but rather understanding what they were doing. It seems obvious that if interviews were used to generate grades in this particular case, the grades would have been different.

Lincoln and Guba (1985), in their remarkable book, related epistemology to research procedures. It is easy to extend such an idea to a relationship between a pedagogy based on a given paradigm of knowledge and assessment. We want to build on this, however, and bring technology into it. A view of knowledge that emphasizes the role of media, symbolized by the humans-with-media unit, must be in resonance with these other components if we are to develop coherent research and practice.

It is therefore important that we keep the discussion about pedagogy and computer technology together with epistemology and research methods. In this chapter, we will borrow the notion of resonance to weave together epistemology, technology and pedagogy, bringing research methodology into the weaving later on in the book.

We agree with Lincoln and Guba (1985) and other authors, such as Bicudo (1979), who stress the relative and intersubjective nature of knowledge. Phenomenologists such as Schutz (see Wagner, 1979) and Bicudo (1979) have emphasized the oddness of framing the debate in terms of which comes first, humans or the world, arguing that there is no sense in thinking of the world without humans beings, nor vice versa. The world could not be thought of as such without humans; on the other hand, it is inconceivable to think of humans without a world. So they argue that 'beings-in-the-world' is the basic and indivisible unit. They propose that humans should be seen as 'beings-in-the-world-with-others', as humans are always relating to others, humans or otherwise. We find this view to be compatible with our notion of humans-with-media, especially when we consider that 'others' in our case, due to the research interest, are the media. In this sense, we can say that we do not believe in a notion of knowledge that separates humans from media, nor knowers from the known.

After representing our view of knowledge, we will answer, in this chapter, the question of why modeling is in resonance with this view of knowledge where media have a fundamental role. In order to do this, we will discuss in the following sections the roots of modeling and relate it to problem solving and problem posing, two important trends in mathematics education in the last two decades. After contrasting modeling with these other two movements, we will present the roots of modeling in Brazil and Denmark, two of the countries in which it is most popular. After that, we will sketch different views of modeling in mathematics education, how they 'interact' with different views of the use of computers in mathematics education, and finally we will point out the limitations of modeling as a pedagogical approach.

2.   PROBLEM SOLVING , PROBLEM POSING AND MODELING

Since the work of Polya (1945), the activity of problem solving has been stressed as very important in mathematics education, although mathematical problems have always been present in mathematical curricula. Polya describes some heuristics and phases of the activity of mathematical problem solving as an exercise of introspection in his own mathematical activity. According to Schoenfeld (1992), "Simply put, *How to Solve It* (1945) planted the seeds of the problem-solving 'movement' that flowered in the 1980's" (p. 352). After the failure of the 'new math' in the 1960's and the back-to-basics movement by the end of 1970's, as a reaction against them, "the pendulum began to swing in the opposite direction, toward 'problem solving' " (Schoenfeld, 1992, p. 336).

Questions about what is considered a problem in mathematics have been discussed. We can find a wide spectrum of 'problems' in books: from routine exercises that require the application of a technique just learned, to perplexing or difficult questions, or even the so-called real problems. In spite of a lack of agreement about what a problem may be, or the fact that discussion about it, was sometimes lacking, this movement gained momentum. Moreover, problem solving has been seen as a means to attain other goals, as a topic to be taught (learning objective), or even as a methodology for teaching.

In an article about historical perspectives in problem solving in mathematics curriculum, Stanic and Kilpatrick (1989) refer to three traditional themes regarding the role of problem solving in the school mathematics curriculum: problem solving as context, problem solving as skill, and problem solving as art. In the first case, various types of problems are used as a means to achieve other curricular aims. In the second case, problem solving is seen as an aim in itself. In the latter case, a trivialization and an algorithmization of the four phases[10] presented by Polya (1945) has frequently occurred, considering them as linear steps or procedures to follow in order to solve a given problem. Problem solving as art, the third perspective Stanic and Kilpatrick (1989) referred to, is directly associated with Polya's work. In this case, the activity of problem solving is considered to be the heart of mathematical activity, but just when non-routine, challenging problems are used. According to Polya (1945), problem solving is a practical skill, like swimming, that is acquired through imitation and practice. But this practice implies the engagement of students in activities that make them live the mathematician's culture, to experience mathematical discovery.

In a Brazilian masters thesis from UNESP - Rio Claro, presented by Gazire in 1988, we found three perspectives related to problem solving in mathematics education that are similar to the ones presented by Stanic and Kilpatrick (1989). Gazire talks about problem solving as a way to apply a mathematical content, as a new content, or as a means to teach mathematics. It seems to us that there is a certain correlation between these perspectives and those previously presented.

Problem solving as a way to apply mathematical content is wide spread in mathematics education and is based on the assumption that students learn a content better when it is applied. In this case, the selection and teaching of mathematical contents and necessary techniques precede problem solving activities and are imposed by the teacher. This approach resembles the first theme identified by Stanic and Kilpatrick (1989): problem solving as a context, since the focus is still on contents. The perspective of problem solving as a new content is closely related to Stanic and Kilpatrick's (1989) problem posing as skill. In this case, it is assumed that the knowledge of problem solving techniques and strategies will help in the development of the problem solving skill.

Problem solving as a means to teach mathematics, the third perspective presented by Gazire (1988), has a more pedagogical tone. The mathematical contents are approached through a challenging, concrete problematic situation initially posed by the teacher. The teacher does not lecture about nor explain mathematics, but guides the students to seek the mathematical

---

[10] The four phases presented by Polya are: 1) Understanding the problem, 2) Devising a plan, 3) Carrying out the plan, 4) Looking back.

contents to solve the problem, analyzing the solutions and evaluating alternative ways to arrive at them. We feel that this perspective is compatible with the vision of problem solving as art, the third theme Stanic and Kilpatrick (1989) associated with problem solving in the classroom. Although Gazire (1988) does not explicitly relate her perspective to the work of Polya, as did Stanic and Kilpatrick, we find some similarities between them, specifically when the Brazilian author refers to the role of the teacher. For Stanic and Kilpatrick (1989), "In Polya's formulation, the teacher is the key" (p. 16) and they recognize that problem solving as art:

> ... is the most problematic theme because it is the most difficult to operationalize in textbooks and classrooms. The problem for mathematics educators who believe that problem solving is an art form is how to develop this artistic ability in students. (p. 17)

The activity of problem solving as Polya proposes is demanding for teachers as well as for students. As Schoenfeld (1992) asserts:

> Even with good materials ... the task of teaching heuristics with the goal of developing the kind of flexible skills Polya describes is a sometimes daunting task. (p. 354)

In this sense, Davis and Hersh (1985), after analyzing an example that shows the heuristic of Polya in action, wonder:

> Do these ideas work in the classroom? The assessment of attempts to transform Polya's ideas into pedagogical practice is difficult to interpret. Apparently there is more to teaching than the good idea of a teacher. (p. 327)

All these authors, Stanic and Kilpatrick (1989), Schoenfeld (1992) and Davis and Hersh (1985), stressed that it is not easy to implement problem solving in the classroom, an assertion that has certainly not impeded attempts to do so. Schoenfeld (1987) himself taught courses in problem solving at the college level, in which one of the goals was the development of executive and metacognitive skills. In order to attain those goals, he developed several techniques with varying levels of intervention of the teacher and interaction between teacher and students. The research that Schoenfeld conducted in the contexts of these courses shed light on the processes the students follow when solving mathematical problems and the way those processes may be improved. We can say that Schoenfeld's courses are compatible with the perspective of problem solving as art, closely related to Polya's work.

In courses that differed from those conducted by Schoenfeld, Gazire (1988) identified three main movements within the perspective of problem solving as a means to teach mathematics in Brazil: ethnomathematics, mathematical models, and the one she has called cognitive-constructivist. She asserted that, in all three movements, the activities begin with concrete situations, and she then proceeded to develop a didactical proposal within the so-called cognitive-constructivist movement. Although she does not talk about modeling, but rather mathematical models, her perspective of problem solving as a means to teach mathematics certainly agrees with the perspective of modeling as a pedagogical strategy we propose. But she did not discuss the kind of concrete situation or problems associated with mathematical models.

The trend to relate mathematics with the real world brought problems of applied mathematics and, consequently, the study of mathematical models, into the sphere of mathematics education. The activity of mathematical modeling developed by mathematicians inspired a new trend in mathematics education. According to Kaiser-Messmer (1991):

> The consideration of applications and modelling examples in mathematics teaching has once again become more prominent since the beginning of the seventies, mainly as a reaction to the almost complete displacement of applications by the structure-oriented mathematics education discussion of the late sixties. (p. 83)

It was a step forward towards bringing modeling, as a pedagogical strategy, into the field of mathematics education.

For Blum (1991), the process of modeling or model building is a part of the process of problem solving. He states: "The applied problem solving process consists of the entire process of dealing with an applied problem in attempting to solve it" (p. 11). Blum described that process as entailing several steps. The starting point is a real problematic situation (outside of mathematics). The first step is to create a real model, making simplifications, idealizations, establishing conditions and assumptions, but respecting the original situation. In the second step, the real model is mathematized, to get a mathematical model. The third step implies the selection of suitable mathematical methods and working within mathematics in order to get some mathematical results. In the fourth step, these results are interpreted for and translated into the real situation. This means validating the mathematical model and may imply modifications. According to Blum (1991), the process of modeling consists of the first and second steps of this problem solving process.

If we now look at the description of the modeling process made by Bassanezi (2002), one of the pioneers of modeling in Brazil, some differences become apparent, since all the steps described by Blum are part of the modeling process for Bassanezi. According to the latter, the steps in a process of mathematical modeling are: experimentation, abstraction, resolution, validation and modification. We can establish a correspondence between Blum's description and Bassanezi's, but there is an essential difference: the appearance of a new element, experimentation, as being a fundamental part of the mathematical modeling process.

At the present time, there are several approaches related to modeling as a pedagogical approach, and we would like to emphasize modeling in the Brazilian perspective. The next section is devoted to the roots of modeling in Brazil, but we would like to anticipate some of their characteristics in order to establish some links and differences with the activities of problem solving, and especially problem posing.

Talking about modeling as a teaching-learning strategy, Bassanezi (1994), states:

> In terms of teaching, the use of modelling leads to the learning of mathematical contents connected to other forms of knowledge. ...
>
> Our job has been to try to connect teaching experience with the modelling perspective based on specific theoretical, philosophical, and methodological concerns. We take into account the human resources, the interests shared by teacher, student and community; social, political, and economical contexts, etc. We also look for the redemption of ethnomathematics, its interpretation and contribution at the level of mathematical systematization.
>
> Working with mathematical modelling does not simply attempt to widen knowledge but to develop a particular way of thinking and acting: producing knowledge, putting together abstractions and formalizations, interconnected to phenomena and empirical processes considered as problematic situations. (p. 31)

In this description, some characteristics of the modeling process can be noted that imply a step forward compared to the perspectives of problem solving described previously: it offers a natural, non-internalist way of connecting mathematics with other sciences; it shows an experimental way of doing mathematics; the mathematical contents connected to the problems arise naturally, as it has an engagement with the problems of the community to which the students and teachers belong; students have the chance of selecting a topic related to their concerns or interests.

We would like to go deeper in our analysis of this last characteristic of our perspective of modeling. The fact that the student chooses a topic to learn about is related to the activity known as problem posing, which is associated with problem solving, but has received less attention from mathematics educators. Although it is recognized as paramount within the field of mathematics, little research has focused on the cognitive processes or classroom tasks related to it. One of the few exceptions is Silver (1994), who refers to mathematical problem posing, stating that: "Problem posing refers to both the generation of problems and the re-formulation of given problems" (p. 19). It is closely associated with problem solving and can occur before, during, or after solving a mathematical problem. Various authors have studied mathematical problem posing that refers to both the generation of problems and the reformulation of given problems. Silver (1994), one of the leaders of this movement, points out several aspects related to problem posing, presenting an exhaustive review of the literature in the area. He states that "Students are rarely, if ever, given opportunities to pose in some public way their own mathematics problems" (p. 19). In the same way Kilpatrick (1987) stated:

> The experience of discovering and creating one's own mathematics problem ought to be part of every student's education. Instead it is an experience few students have today– perhaps only if they are candidates for advanced degrees in mathematics. (p. 123)

This is a characteristic of the traditional model of mathematical instruction that still persists in school, where the teacher speaks and proposes problems, generally from a textbook, and the students listen and try to solve those problems. In this way, the problem posers are teachers or textbook authors, but not the students. Some examples of problem posing tasks we have found in the literature will help us illustrate the issues we would like to emphasize and to establish some differences between problem posing and modeling as pedagogical approaches.

The task in Figure 3-2 corresponds to the first part of *The billiard ball mathematical (BBM) task* of an exploratory study conducted by Silver, Mamona, Leung, and Kenney (1996) with middle school teachers and prospective secondary school teachers. The second part of the task is a problem solving one, and the third part asks to pose new problems (see Figure 3-3).

Silver, Mamona, Leung, and Kenney (1996) assert that problem posing as a focus of attention in mathematics education may occur: 1) prior to any problem solving (generating problems from an invented or naturalistic situation); or 2) after solving a given problem (generating new problems inspired by the original one). In the tasks presented in Figures 3-2 and 3-3, the authors tried to elicit problem posing activities at both moments, before and after solving a problem.

(Part 1)
Imagine billiard tables like the ones shown below. Suppose a ball is shot at a 45° angle from the lower left corner (A) of the table. When the ball hits a side on the table, it bounces off at a 45° angle.

In each of the examples shown below, the ball hits the sides several times and eventually lands in a corner pocket. In Example 1, the ball travels on a 6-by-4 table and ends up in pocket D, after 3 hits on the sides. In Example 2, the ball travels on 4-by-2 table and ends up in pocket B, after 1 hit on the sides.

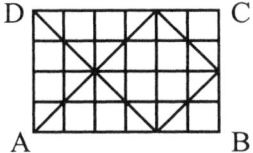 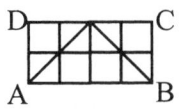

Look at the examples, think about the situation for tables of other sizes, and write down any questions or problems that occur to you.

Figure 3-2. The first part of the BBM task (Silver et al, 1996, p. 297)

(Part 2)
Look at the examples, think about the situation for tables of other sizes, consider as many examples as you need, and try to predict the final destination of the ball. That is, when will the ball land in pocket A? When will it land in pocket B? In pocket C? In pocket D?

(Part 3)
As you work out your solution to the problem, other questions may also come to mind. In the space provided below, write down any questions or problems that occur to you.

Figure 3-3. The second and third parts of the BBM task (Silver et al, 1996, p. 297)

Another kind of problem posing task is supplied by Silver and Burkett (1994). They studied the posing of division problems made by pre-service

elementary school teachers from a public university using the task in Figure
3-4.

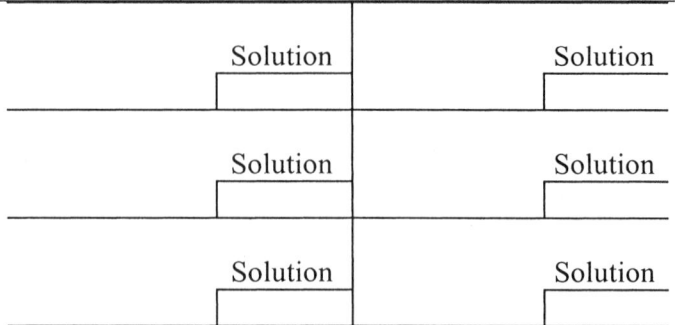

Please work individually on the task given below.

$$40 \overline{\smash{)}540} \quad \begin{array}{r} 13 \\ \end{array}$$
$$\underline{40}$$
$$140$$
$$\underline{120}$$
$$20$$

In the spaces provided, write as many different story
problems as you can that match the computation shown
above. The story problems you propose must all have
different solutions. For each story problem you propose,
indicate the solution in the box provided.

Solution	Solution
Solution	Solution
Solution	Solution

Figure 3-4. A task for pre-service elementary school teachers (Silver and Burkett, 1994)

In this case, the researchers reported, among other results, that about 25%
of the problems posed had some unreasonable conditions, showing weak
connections between mathematics and the situation presented in the
problem. It seems to us that the problem posers create an artificial situation
and a problem related to it, but they do not verify if the mathematical
operation and its result are coherent in that context. This situation shows the
disconnection between mathematics and real situations that often happens in
school.

We would also like to present some examples of problem posing tasks
given to children. English's (1998) paper presents third-grade children's
problem posing within formal and informal contexts. In formal contexts, for
example, a large card displaying an operation like 12 - 8 = 4 was shown to
the child, and then the researcher asked: "See if you can make up a story
problem that could be solved by this number sentence". In the informal

context, symbolic mathematical representations were avoided; and pictures, stories from literature or non-goal-specific statements were presented to the children. An example of such statements is: "Sarah has five dolls on one shelf in her room and four toy cars on another shelf". After showing the picture or telling a story, the child was asked: "Can you make up a story problem that asks about something you can see in the photograph?" or "Can you make this into a problem that we could solve?"

In a similar way, Silver and Cai (1996) aimed to analyze arithmetic problems posed by sixth and seventh-grade middle school students, and they presented the task in Figure 3-5

Write three different questions that can be answered from the information below.

Jerome, Elliot, and Arturo took turns driving home from a trip. Arturo drove 80 miles more than Elliot. Elliot drove twice as many miles as Jerome. Jerome drove 50 miles.

Figure 3-5. The problem posing task (Silver and Cai, 1996, p. 525)

It is remarkable that, in this case, about 25% of the children's responses were considered by the researchers as non-mathematical questions and of marginal interest for the study, because they were interested only in mathematical tasks. Nevertheless, the authors considered that these data may constitute a potentially interesting source for other issues related to mathematical problem posing. The novelty of asking students to pose 'questions', and the ambiguity of the task, were some of the reasons the researchers cited as possible explanations for those responses. The authors reported some answers that they considered to be related to students' personal commitments and values (morality, justice, human relationships) when some students revealed an apparent concern with the fair distribution of driving responsibilities: "If they each drive an equal amount, how many miles would each person drive?" "Why does Arturo drive so long?" "Why did Elliot drive twice as far as Jerome?". The authors finally assert:

Although it is not possible to know precisely the underlying reasons for these unexpected responses, their appearance suggests that an open-ended problem posing task, which invites students to express their own questions, may lead to outcomes (i.e., posed problems or questions) different from the ones a teacher or researcher might have in mind. (p. 537)

In all of the previous examples, the answers of the subjects (teachers or students) were conditioned by the situations presented by the researchers. Non-mathematical questions appeared that were treated by the researchers as being of marginal interest (Silver and Cai, 1996), but those questions indicate that the students' interests may transcend mathematical issues. The problems the researchers are expecting from the problem posing tasks are mathematical problems, and thus, the other questions were considered marginal. This interpretation suggests an *internalist* characteristic of the usual problem posing tasks in mathematics education. The artificiality of the situations that the students have to create leads to some incoherence when trying to connect mathematics and real situations (Silver and Burkett, 1994). Some of the studies reviewed by Silver (1994) found weak connections between real life situations and mathematical ideas or symbols among students or teachers engaged in activities of problem posing. The numbers selected by many of them were incoherent within the context of the problems, and the level of difficulty of the problems they posed was more related to computational complexity than to situational or semantic complexity.

It is not our intention to analyze the kind of problems posed by teachers or students in answer to the tasks discussed here; we would simply like to contrast these problem posing activities with the ones performed in an educational context where modeling is used as a pedagogical approach. Problem posing within modeling goes beyond the fact of generating or formulating mathematical problems. In our perspective of modeling, the students have to first choose a topic to study, which may be outside of the mathematical realm, and then propose problems related to that topic, but again those problems may not be mathematical.

Silver (1994) considers that problem posing is a natural feature of inquiry-oriented pedagogy where students are encouraged to be autonomous learners, and mathematics can be empowering for them. This view of problem posing in conjunction with an inquiry-oriented pedagogy is closely related to the modeling approach as a pedagogical strategy that we are proposing and, at the same time, constitutes a challenge to traditional mathematical instruction. But we can consider the students' reactions to activities of problem posing in mathematical classrooms. Silver (1994) refers to problem posing as a means of improving students' attitudes and dispositions toward mathematics. This aspect shows two opposite faces. On one side there are students who engage positively in that activity, stimulating their interest in mathematics. On the other side there are students who have been successful in environments characterized by teacher-directed instruction, who resist changes and react negatively "to deal[ing] with higher level of uncertainty about expectations or higher level of responsibility for their own learning" (Silver, 1994, p. 25). In this respect, our experience with the modeling approach in a mathematics course for biology majors shows the initial difficulties of students in choosing a topic they would like to learn about. They experience surprise and disorientation in the face of that unexpected activity. In this phase, the ability of the teacher to foresee the potential of the topics in mathematical terms is essential. After the initial impact, and having decided which topic they would like to study, the task of problem posing related to that topic is not a simple one. Sometimes, in spite of the students' open-ended explorations, there are problems that may not arise naturally for them. In this case, the teacher has a paramount role as a guide: suggesting new paths, and references, asking for information from a specialist in the area, or being a problem poser him/herself. Kilpatrick (1987) also refers to the teacher's responsibility for engaging students in creative processes of problem formulation, and stresses collaborative work with other students to improve problem solving abilities, as well as problem posing abilities.

Problem posing is considered as a central feature of mathematical activity: conjecturing, reformulating ill-structured problems, or posing new problems are at the heart of mathematical thinking (Silver, 1994), but it is not exclusively from mathematics. Although the position that it is important to provide students with the opportunity to experience what mathematicians experience is widespread inside the mathematics education community, we believe our concern as educators goes beyond this goal, since in modeling, the students choose a problem that could be outside of what is normally considered to be mathematics and, in this way, the problem posing activities transcend mathematics, entering other realms of science and making the work interdisciplinary. In this sense, the modeling approach is closer to the activity of an applied mathematician. Pollak considers himself an industrial mathematician, and he states:

> In my job, problem formulation means to take a fuzzy, ill-defined situation in some other field, or in the real world, for which there is no obvious mathematical formulation or structure, and formulate a mathematical problem that will help make the situation precise and qualitatively, structurally and analytically comprehensible. (Pollak, 1987, p. 255)

These activities mentioned by Pollak are similar to those developed by students who are engaged in a modeling project. Initially, the students, working in groups, select topics for inquiry according to their interests; they propose problems in that context, that may need future reformulation, and they devise a plan to solve them. Sometimes the mathematical tools they have are sufficient to generate a model that accounts for the situation; other times they have to make mathematical explorations or study some additional mathematical topic. The participation of students is central, and since they are free to select a topic, they consequently share the responsibility for their learning processes. Let us look at some examples of this initial activity of topic selection and problem posing in modeling contexts.

In 1983, the teacher of a beginning calculus course for food science majors, at a public university in Campinas, Brazil (UNICAMP), encountered his students on the first day using t-shirts with the inscription *Detesto Cálculo* (I hate calculus). The students felt the study of calculus to be something without any utility. At the beginning of the course, the teacher proposed that the students work only with the mathematics they felt useful and interesting, using problems or situations they would pose. Various topics arose: optimization and handling of packaging, diets, balanced animal rations, etc. The topic they selected was potato cultivation, a theme a student proposed in the following way: "My father plants potatoes, placing each seed 30 cm apart; I would like to know why he does it this way". There was no immediate answer, and the first step was to get information from the Secretary of Agriculture. After this activity, an initial problem was raised: Determine the spacing between two plants (in the same row) in such a way as to maximize the production in one alqueire[11] (Bassanezi, 2002). Other mathematical problems may have been posed in the development of the students' work while analyzing the data they collected: What kind of function fit the data? How can that function be constructed? Is that function a good model for this situation? In which range is it valid? More details about this example will be discussed in Chapter 6.

In 1986, the Energy Company of São Paulo (*Companhia Energética de São Paulo-CESP*) promoted a campaign to save electrical energy, and the authorities of the schools were asked to collaborate with the campaign by informing the students in the classrooms about it. In one fifth grade mathematical class from a low-income public school, the students became engaged in the campaign, and the teacher decided to develop the mathematical contents that could be associated with that topic. The problem raised was: How to calculate energy consumption? (Gustineli, 1990).

In the 1998 course in Applied Mathematics for biology majors at the State University of São Paulo (UNESP), Brazil, where the teacher used modeling as a pedagogical strategy, a group of students was interested in the behavior of the stomata in different environments. They studied the alterations of the stomata of a certain plant in places with variable light, humidity and pollution levels. They selected the variables and collected data,

---

[11] *Alqueire* is a unit of area used in some parts of rural Brazil, and it is equivalent to 24200 m^2

and then made tables and graphs in order to decide what kind of function could fit the data (Borba and Bovo, 2002).

The kinds of topics that appear when modeling is used as a pedagogical strategy are sometimes unexpected for the teacher. In the case of college-level mathematics courses for non mathematics majors, the topics raised by the students are generally associated with other disciplines. Nevertheless, mathematics topics have also arisen in different educational levels: for example, a group of high school students in the U.S.A. decided to study fractals (Borba, 1993a), and this topic was also proposed in 2001 by a group of biology students at a Brazilian public university who were enrolled in a mathematics course where modeling was used as a pedagogical strategy, just because they were curious about that topic.

All the examples that involve posing of problems in the modeling perspective have a clear tendency to be interdisciplinary, and they show a non-internalist view of mathematics. Also clear is the challenge to the classical model in mathematics teaching, which can be characterized by the sequence: lecture-examples-exercises. In such a model, we can observe a frequent inclination to connect mathematics with the real world through the exhibition of some applications during the lecture followed by the solution of problems, applying the mathematical contents recently taught by the teacher. In the modeling approach, the situations proposed by the students provide the opportunity to pose and solve mathematical problems naturally considering the limitations coming from the real situation they have chosen to investigate. The applications are not artificial; they precede the mathematical contents and make plenty of sense to the students.

There are papers that report on a strong relationship between mathematical knowledge of the study participants and the quality of the problems they posed, considering problem posing to be a feature of creative activity or exceptional mathematical ability. In spite of these assertions, we think that, although exceptional mathematical ability would be of great help for the processes of problem posing, there is no need to consider the complexity of the problem posed as a measure of mathematical giftedness. We strongly believe that those activities are a right of the students if our goal is to encourage autonomy, participation, collaboration and democracy as paramount values of citizenship that the school ought to develop.

At this point, other differences between problem posing within the modeling approach and problem posing as a partner of problem solving as presented in the literature might be pointed out. Despite the didactical value of problem posing in mathematics education, we believe that problem posing, in the context of a modeling approach, also has a socio-political value, since the basic guidelines of modeling, in the Brazilian perspective, are, among others:

> To take into account the specific realities of every region and the students' interests, aiming at increased motivation and at an effective participation of the students in their communities or in a larger context in which they take part. (Bassanezi, 1994, p. 31)

and

> To appreciate the human resources, explore and develop teachers' and students' skills, making them feel able to give the community their contribution and form socially active individuals. (Bassanezi, 1994, p. 31)

These aspects are strongly related to our responsibility as educators to form democratic and participatory citizens in their communities, and have been absent in the discussion about the importance of problem posing in mathematics. Little attention has also been given in the literature to the conjunction between problem posing and technology, with the exception of Kilpatrick (1987), who remarks on the role of the computer as a support for exploratory activities that may generate problems and yield conjectures that students can test and try to prove. The association of modeling with computers is fertile ground for the kind of mathematical problem posing that takes place in modeling.

In summary, both problem solving and problem posing represent a step forward compared to traditional classroom activities involving the use of textbooks, in which the order is commonly 'theory, example and exercise', an approach well-criticized by Baldino (1999). However problem solving and problem posing, for the most part, seem to be connected to an internalist view of mathematics as opposed to an interdisciplinary view that characterizes the perspective of modeling we are presenting. In both trends analyzed, students have no role in designing the curricula or topics to be studied, whereas in modeling, they do have a role, as they suggest a theme to be studied and negotiate how the problem will be defined. By the very nature of it, students have to deal with mathematics not only developed by mathematicians, but also by physicists, chemists, carpenters, and whatever community they are lead to.

Another contrast we would like to emphasize is the one connected to social-political aspects. By having students choosing themes, the door is open for them to bring with them part of their cultural background and their concerns with the future, and for political themes to enter the mathematics classroom. But there is also another issue that cannot go without mentioning. The act of choosing the problem and negotiating this problem with the teacher suggests an aspect of the hidden curriculum: students have to take on a new role of not answering to assignments. This is what makes the posing of the problem significantly different in modeling compared to problem posing.

As mentioned before, computer and information technology is another issue that distinguishes problem posing/solving from modeling. Computers, with good reason, have remained distant from problem solving and also from problem posing. In contrast, modeling in mathematics is continuously supported by the development of computer technology, and as we will see later, computers have also changed the face of modeling in mathematics education.

Now that we believe we have framed modeling within these other two trends in mathematics education, we will present the way modeling has been shaped both in Brazil and in Denmark, countries in which it may not be the pedagogical approach with the highest status, but where it is at least recognized as important.

3.   ROOTS OF MODELING IN BRAZIL

There are several definitions of modeling in Brazil. Many considered it to be the process of repeating, in the classroom, the process that applied mathematicians go through when they create new models to describe a given situation and predict outcomes of an ongoing process. Others consider modeling in mathematics education to be characterized by the use of long-term problems in the classroom, and some others stress the interdisciplinary possibilities of such an approach.

We would like to think of mathematical modeling as a metaphor for a pedagogical approach which stresses the students' choice of the problem. We, ourselves, like other authors in Brazil[12], draw on certain aspects of mathematical modeling in order to propose pedagogical approaches within mathematics education. The aspect we emphasize is that problems have to be shaped and defined by the group that is facing a particular obstacle. It is also true that, for the most part, mathematical modeling is developed by groups of specialists from different areas of knowledge, and often the work they develop is interdisciplinary. With these two aspects of modeling - students' choice of the problem, and interdisciplinarity - we want to challenge the basic structure of curricula in most places of the world: curricula is

---

[12] Although the second author of this book is Argentinean, we refer to her here as if she were from Brazil for two reasons: first, because she lived in Brazil for 4 years and has visited the country regularly due to professional activities; secondly, because modeling is not yet a trend in Argentina

organized in disciplines which resemble the way science departments were classified, and put into a hierarchy, during the 20th century.

In our understanding of modeling as a pedagogical approach, which is also named *project work* in Denmark, the order to be followed will be that which interests the students. In this sense, teachers or professors become co-authors of an investigation developed by the students, and will often bring the more established scientific perspective into the group. It is very likely that the interest of a group will not coincide with the official boundaries of a given discipline. In the case of mathematics, it means that teachers will have to be more open to venture into other fields, as well, and/or help the group to find specialists in the university or in homepages that can provide some information on a given theme.

A paramount issue in such a pedagogical approach is, therefore, the choice of the problem. Just like research in applied mathematics or any other field, finding the problem, and defining a piece to be investigated, is the hardest part of a scientific endeavor, and often such a choice will decide the success of a given research group.

Choosing a problem, as opposed to an imposed or pseudo-problem (Borba 1987), is rooted in Brazil in the work of Freire (1992). One of the underpinnings of his famous literacy method was that both teacher and students would bring words (themes) of their interest into the literacy process. As a person ahead of his time, Freire was already breaking barriers between the disciplines, as it was not possible to know whether, in the method he developed, someone was learning how to write, or was learning social science or ethics. Similarly, when modeling can really penetrate a classroom atmosphere, and students interact with the teacher in order to choose a theme to study, students sometimes do not know whether they are studying biology, politics or mathematics. In both cases, students are also having their cultural input taken into account - by the choices of 'words' in the case of literacy, or the choice of the theme in the case of modeling.

Freire's pedagogy was also acknowledging that different ways of knowing could enter the classroom. In the search for words and themes, adult students who were learning how to read, write, 'read reality' and be citizens were also bringing their stories and their type of knowledge into the classroom. In doing so, they were challenging the presence of just one type of knowledge in the classroom: the scientific one. This educational movement was also showing the world the connection between culture and learning, since the problem chosen by the learner is embedded in the culture they are actively a part of (Borba, 1987).

Similarly to Freire, D'Ambrosio (1985) brings diversity to the mathematics classroom. He takes pains to point out that mathematics is also culturally bounded, and that it is necessary to understand that what we have called mathematics is just one form of expression. In his most oft-quoted paper, D'Ambrosio (1985) conceives of ethnomathematics as:

...the mathematics which is practiced among identifiable cultural groups, such as national-tribal societies, labor groups, children of a certain age bracket, professional classes and so on. Its identity depends largely on focuses of interest, on motivation, and on certain codes and jargons which do not belong to the realm of academic mathematics. (p. 45)

With this publication, the mathematics education community became familiar with a notion that he and some of his collaborators had been developing for over a decade by that time. D'Ambrosio has, over the last 25 years, clarified and modified the notion of ethnomathematics as part of a movement that has grown in Brazil and elsewhere as well. In the late 1970's, when the notion of ethnomathematics was being elaborated by D'Ambrosio (1978), the relevance of models for science and mathematics education was also beginning to be noticed.

For the purpose of this discussion, it matters that D'Ambrosio and Bassanezi proposed, throughout the 1980's, the notion of modeling as a pedagogical approach characterized mainly by the use of problems from the real world, choice of the problem by students, and resemblance with mathematical modeling. As the reader will see, ethnomathematics and modeling had a successful marriage in Brazil, as the choice of a problem by students was associated with the cultural boundaries of the problem itself. As pointed out earlier, this is also the reason why modeling coming from applied mathematics soon gained political and socio-cultural perspectives.

The relationship between ethnomathematics and modeling was emphasized as the Graduate Program of Mathematics Education at UNESP, Rio Claro, São Paulo was started. Although there were other short-term, graduate level courses in Latin America, it can be said that this program was one of the first graduate programs in mathematics education in Latin America, as it started its Masters Program in 1983 and the Doctoral Program in 1992. In this program, ethnomathematics and modeling were the main trends in the 1980's. The first research using the ideas of mathematics educators such as D'Ambrosio and Bassanezi was also developed in this program. The second master thesis of this program was written by the first author of this book in 1987 (Borba, 1987) and had two goals: to map the ethnomathematics developed in a shantytown (*favela*) in Campinas, a city with approximately 770,000 inhabitants at that time, located about an hour north of the city of São Paulo, the capital of the state with the same name; and to study how to incorporate such ethnomathematics into educational practices within the same slum.

In that study, a long term investigation, using a participatory research approach, was carried out. The mathematics of adults who worked, for instance, as plumbers, and the mathematics of the kids who played soccer were documented in that report. Then, the researcher struggled with the problem of how to use Freire's, Bassanezi's and D'Ambrosio's ideas with children, instead of adults. Freire had worked mostly with adults from poor neighborhoods, Bassanezi with university level students, while D'Ambrosio (1978) had worked with ethnomathematics solely at a theoretical level. How to bridge such ideas about citizenship, modeling and culture in a context where kids, slums and informal education were main characteristics?

Kids in the slums in Brazil do not usually have a community center to go to as did these boys and girls. A community association from that shantytown had demanded such a center so that, among other reasons, "their kids would not get involved in petty theft". Once a community center was built, lay teachers were hired by the city government, again in response to the demands of the neighborhood association.

In 1983, a group of students from another state university, UNICAMP, started to develop a pedagogical project within this community. This group included two graduate students in education (supervised by Paulo Freire) and undergraduate math and physics majors. In 1984, only the two graduate students who worked in the area of literacy remained, and Borba joined this group as a masters student from UNESP.

It was in this context that Borba tried to find an answer to the question regarding the use of ethnomathematics in educational practices with kids. The kids developed activities within the community center, which they called 'the little school' (*a escolinha*), and played soccer and other games outside. After the researcher was accepted by the lay teachers and the group of kids, and had a chance to study some of the mathematics developed by members of the shantytown, he asked the kids several times if they wanted to study or research some kind of theme. The kids, whose ages varied from 8 to 13, had many ideas, but the ones that persisted seemed to be connected to their culture.

The challenge, as an action researcher, was to bring changes to an informal educational process in a way that could incorporate the mathematics of plumbers, vegetable gardens, housing and a soccer field. The resonance between modeling and ethnomathematics was part of the solution. On the one hand, Borba could not just impose that kids would study the mathematics known by other kids or their parents, and on the other hand, he wanted to value their culture as a means of building citizens, which also means kids with self-esteem. The emphasis on having students choose themes for their investigation had many advantages, among which we will emphasize the following: different kids could be dealing with themes which could accommodate diverse interests; boys and girls could express, in their choice of the problem, their different cultural backgrounds; and the lay teachers who were responsible for them could deal more easily with students who were not so accustomed to following teachers in structured activities.

One of the themes, which actually merged two others, was the vegetable garden and the sale of its products. In one group, the children developed the notion of scale, which was unfamiliar to them, and they designed a chart to organize the garden. Another group studied arithmetical concepts and skills that allowed them to do the bookkeeping correctly.

For the purpose of this book, the above examples from the slum illustrate that the roots of modeling in Brazil were related to the notion of ethnomathematics. As many authors express (Bassanezzi, 1994), the interplay between these notions is promising, as curriculum development can take into account and value different cultural backgrounds. In this sense, an ethnographic study about mathematical practices of a given socio-cultural group may be important for teachers so that they can help students in their modeling activities. Bassanezi (1994) and Borba (1999a) have stressed how technology can be another important partner in such an approach. As students are developing a given project, based on the theme they chose with the guidance of teachers, students can, for instance, use computer technology to help with numerical analysis (Bassanezi, 1994) or with curve fit (Borba, 1999a). Computer technology then becomes an important co-actor in order to prevent a given investigation from losing its momentum due to long detours with paper and pencil.

At this point, the reader may be convinced that modeling in Brazil has a strong social component, as its association with ethnomathematics led many researchers to become involved in underdeveloped rural areas, workers' and peasant movements, and so on. But modeling is also practiced in regular public universities, as we will see in later chapters, and is no longer connected only to slums and peasants as it was in the 1980's (although such connections do persist). This political concern associated which the notion of the students' choice of the problem is also characteristic of the work of Dewey, who developed his work early in the last century in the U.S.A. and influenced many educators in Brazil, as well. Experiences of the students were valued in Dewey's proposals for curricula. Dewey may be in a process of being rediscovered, both in Brazil and in the U.S.A., partly because the notion of projects and experience are intertwined. His work in the U.S.A., however, seems to be confined, in practical terms, to the area of literacy and to a few schools in the southern U.S.A. that try to implement some of his ideas in formal education. As Dewey had some influence in Brazilian education in the first half of the 20th century, it would not be surprising if he were one of the sources for these different movements in Brazil, including the one we have been calling the modeling approach, but this remains to be established.

In any case, the work of neither Dewey nor Freire took into account the role of computer technology. In the case of Dewey, the explanation is obvious; in the case of Freire, it can be argued that he would have discussed it in one of his books if he had not died in 1997, since during his term as Secretary of Education of the city of São Paulo in the late 1980's, he sponsored a large program to establish computer laboratories in many schools in the poor neighborhoods of the largest city in South America.

Although the influence of Dewey on the Brazilian tradition is not clear, we can say that his ideas regarding experience and education are comparable with some aspects of Freire's proposal for education. Freire has had influence on the work of D'Ambrosio and Bassanezi, who led the modeling movement in Brazil. Modeling in Brazil has developed both as a 'branch' of mathematics and as a pedagogical approach in mathematics education. In Brazil, modeling appears to have this political and interdisciplinary approach, and has been associated with technology. Now let us examine some similarities with the Danish case.

4.   PROJECT WORK: ITS ROOTS IN DENMARK

Niss (1977) is probably the first reference in English about what became a tradition in mathematics education: the Danish project work. The author makes sure that he situates the Dutch and German influence in his work, but it should be said that it seems to be in Denmark that such a perspective became more established. The basis for the above assertion is that, in Denmark, two universities were created around the notion of project work (Vithal, Christiansen and Skovsmose, 1995).

The idea of having students working on long term projects can be seen as a reaction to the traditional ways of teaching, in which mathematics is taught in an internalistic way. Niss (1977) states that "mathematics instruction is in a 'crisis', a crisis of relevance" (p. 303). He points out, based on the work of other researchers as well as his own, that traditional pedagogy based on hierarchical sequences is in crisis, as students fear the subject and "are bored, work either too much, without really getting anywhere, or too little but strained with a constant bad conscience and a permanent feeling of insecurity" (p. 303). He points out that this crisis is occurring in basic education as well as at the college level. The crisis in basic education is intensified by the fact that teacher education is based on the idea that teachers are educated to be mathematicians. As they start to teach, they reproduce their experience in the university, and the crisis starts at the basic level. Niss, in this same paper, argues that such a crisis also became a part of the university, as students felt that they should be prepared to become teachers.

Niss (1977) proposes that a solution for this could be a problem-oriented curricula, which is close to what we are labeling modeling in this book. He supports this view with the relevance argument, since complex, real situations are chosen to be studied. Similar to the Brazilian tradition already discussed, limiting and defining the problem became part of students' activity, although in his report as well as that of Vithal, Christiansen and Skovsmose (1995), the theme, and sometimes even the problem, were still defined by the teachers. In the examples presented in Skovsmose (1994), the themes and the problems are also chosen by the teacher.

In any case, Niss (1977) also bases his argument on the premise that the division between applied and pure mathematics is not so important, and that both should be considered mathematics, not just the latter. Therefore, the starting point for students should be modeling and refinement of the problem until they feel the need and are encouraged by teachers to study given topics of 'pure mathematics' which are important for the project they are developing.

Niss (1977) also points out tensions between the notion of project and being able to address a given topic in mathematics, and suggests that such tension is part of an inherent dialectic process of such a pedagogical proposal. One should be aware of it, but it is never possible to completely overcome it. Vithal, Christiansen and Skovsmose (1995) do not refer to Niss' paper, but they seem to be aware of this tension as they celebrate concepts associated with modeling-problem-centered studies, interdisciplinarity, participant-directed studies, and the exemplarity principle – which are in the guidelines of one of the universities that built its entire curriculum around project work. At the same time, they also make the criticism that some of these concepts have become weaker or lost power as they analyze the situation in practice.

These three authors present the project of the university, in which 50% of the work is connected to projects and 50% to courses. They then present how projects have increasingly approximated applications related to physics instead of open situations which had social impact. The authors are very critical of such developments and assert that part of the origins, which include the uprisings in the universities in the 1960's with their claims for democracy at all levels, were being betrayed by this route that project work was taking in Denmark.

It is fair to say, however, that even if the assessment made by these authors can be extended to other universities in other educational settings, Denmark is probably the country where modeling, as we discuss it in this book, is more institutionalized. Modeling in Denmark also has roots in mathematical modeling, and it has a social perspective which was also influenced by Freire and D'Ambrosio, since Skovsmose (e.g. 1994) started to discuss some of his ideas and connect them to project-oriented curricula in a critical perspective.

Christiansen (1997), who did research about communication and negotiation of meaning among students in a modeling course in Denmark, illustrates that such an approach is still present at the university level. In trips to that country, the first author of this book observed how this pedagogical approach works and how there are already entire schools working in this way.

It is interesting to note that, although both Skovsmose (1994) and Christiansen (1997) deal with examples that involve computer technology, its interface with the modeling pedagogical approach is not considered in their papers. This seems to be a contrast between the references we have analyzed from the Danish tradition and the research we have developed in GPIMEM over the last ten years. After having presented both the Brazilian and the Danish tradition, with their similarities and differences, we will now discuss specifically the relation between information and communication technology and modeling.

5.   MODELING AND INFORMATION AND COMMUNICATION TECHNOLOGY

Early in this chapter, we argued that modeling as a pedagogical approach is in resonance (Lincoln and Guba, 1985) with the perspective we adopted in this book regarding information and communication technology. Afterwards, we framed modeling, contrasting it with other trends in mathematics education such as problem solving and problem posing, which have differences and similarities with modeling. But as is common in this 'post-modern' scenario we live in today, very few terms or trends have just one single definition or conception. This is also the case with modeling, and with perspectives regarding the use of computers in (mathematics) education in general.

Let us analyze the case of modeling first. If we consider the thorough review made by Barbosa (2001), there are many different ways in which modeling could be classified, and these means of classification generate an even greater range of variations of what different authors understand by modeling. All of them, however, are, one way or another, connected to the very roots of modeling: applied mathematics. Practically every author out of the approximately 100 references analyzed by Barbosa (2001), indicated such an origin. It is possible to observe that, for some, applied mathematics is the beginning and the end of the story, as they claim that modeling in education involves application to 'real problems', and no more is said. A second perspective, that is very connected to this common origin, relates modeling to a given topic of mathematical content that has already been developed. For instance, modeling would be restricted to activities that would lead to the use of some kind of differential equations. A third perspective regards the use of projects in which a theme is chosen by the teacher and the teacher proposes a problem in the form of a question. Finally, there are perspectives that more closely approximate those proposed in this book, in which no theme or problem is given a priori; students are invited to choose a theme, and, with the help of the teacher, they set up a problem to be investigated.

As mentioned before, there could be other ways of classifying modeling, and the distinctions made above do not cover all the possibilities which have already been put into practice. It is important, however, for the discussion regarding the relation between the modeling approach and the use of technology in mathematics education, to note that if modeling is defined too broadly, we run the risk of including even those common math problems given at the beginning of each chapter of traditional textbooks. In the second perspective, if the main criteria to define whether modeling is taking place or not is the presence of a mathematics topic, one loses the perspective that modeling can loosen up the structure of disciplines that schools inherited from the way science has come to be organized over the past two centuries.

Finally, the distinction between the third and the fourth perspective discussed above has to do with the control of the topic to be studied. It is characteristic of the current school system in which themes, problems, and assigned exercises are all chosen either by the teacher, by the author of the text book, or by school administrators. For the purpose of this book, we will refer to the teacher as being the actor who controls such a choice in the classroom, even though we are aware of the complexity of the problem. In any case, as the choice of the theme and of the problem is still made by the teacher in the third perspective, a project is already placed within boundaries. Moreover, often there is already a mathematics topic planned to be explored within the given theme, so in this case, there is less of a rupture with the structure of curriculum as it is currently organized.

In the fourth perspective, students choose the theme and have a strong voice in shaping and designing the problem to be investigated. Of course, the teacher helps the students to define the problem and to make sure the problem is feasible within the structure of the educational institution they are in. Therefore, in this perspective, the choice of the student is also shaped by cultural issues as a whole, and in particular by the way the school is organized, the way the library works, and the design of the Web pages used by the students. But what is important to observe is that such a possibility breaks free from a 'sacred cow rule' in which students have little or no say in curricula. As we have seen and will continue seeing further on in this section, this has political implications, and it also has synergy with different perspectives of the use of computers in mathematics education. Let us now analyze some perspectives regarding the use of technology.

Computers can be seen as tutors. In this perspective, computers attempt to 'substitute' teachers, and the role of the teacher is seen as the one who answers questions. Databases are constructed either with classic programming techniques or using 'neural nets' concepts in which the feedback provided by the computer is adjusted to the history of answers given previously. Such an approach has been used widely to prepare for exams in which memorization is essential. The idea of providing courses through the Internet to large audiences is also very popular among followers of such a view of the role of computers in mathematics education. One can reach a wide audience, with little interaction, generating more profit.

Many see computers as a means of motivating their students. In this perspective, the main reason to adopt computers in education is because "it is more fun", and it allows students to always be motivated to study the topics that the school feels must be taught. There is very little research, if any, backing this kind of claim, which is very popular among teachers and researchers when they are asked about reasons to use computers in education. In our experience, interest in a new software or device varies among students and tends to dissipate relatively quickly. Therefore if motivation were the main argument for having computers in school, large sums of funds would be needed to keep students motivated with new software, machines and interfaces throughout a semester-long course, to say nothing of longer periods.

Neither perspective about the use of technology summarized above emphasizes epistemological issues, so knowledge production is not the focus, or is assumed to be untouched by the presence of new technological actors. In contrast, computers, graphing calculators and different interfaces linked to them can be seen as reorganizing thinking and altering the nature of knowledge production as collectives of humans-with-media are altered qualitatively by the entrance of new members. Computers should be in school, according to this train of thought, because students should be exposed to these new technologies of intelligence so that the knowledge which is produced in schools and universities is not disconnected from the rest of society, since information and communication technologies are an increasingly integral part of our lives.

In our opinion, the main reason for promoting the use of computer technology can be identified by the word 'citizenship'. According to this perspective (Borba, 2002), access to computer technology is not necessarily connected to epistemological issues, but is basically a 'right' that citizens should have in their education. Literacy, matheracy and technoracy are seen at the same level.

As the reader may have deduced, we hold positions that are based on the last two arguments. For the purposes of this chapter, however, what we wish to emphasize is that both modeling and computers have a wide range of perspectives, and that if one considers the different combinations within the classification of modeling developed in this section, and the perspectives of technology we described, it is possible to construct a matrix in which its 'cells' would be the results of the interaction between different perspectives in modeling and technology.

Table 3-1 illustrates this idea. For instance, if one holds the perspective of computers as a tutor, and sees modeling as any kind of applied problem found in traditional textbooks, very few changes may have to be made in the classroom. The combination of both approaches can result in an 'electronic book' in which regular texts of a textbook are scanned into digital figures and students get feedback regarding the correctness of their answer or not. In our opinion, the computer medium becomes domesticated in this case[13], as it is taking very little advantage of the possibilities of this medium as it tries to reproduce the way written texts are presented. In this case, by analogy, one may say that modeling is also being domesticated.

Another potential 'resonance' among views of technology and modeling might be represented by the cell that characterizes technology as "motivation" and modeling as a "mathematical topic". Let us say that modeling is defined as an activity that involves differential equations. Sites could be designed to have many situations in which such a concept would be used. Links could be made available to homepages, in which specific techniques to solve given kinds of equations would be presented, and to software that could 'solve' these kinds of equations analytically or numerically. As mentioned before, a huge amount of work in this site would be necessary so that the motivation would not vanish after a few sessions. If we consider the bottom cell in the motivation column, we could imagine a situation where the teacher adopts the fourth, more participatory version of modeling, inviting the students to choose a theme of study (and thus contribute to the construction of curriculum), but the students are motivated only to use the Internet and search programs. Of course, depending on what

---

[13] See Chapter 2 for more on this issue, in particular the section entitled "Media, humans and knowledge: possibilities of merging".

part of the world the students are in, what social class they belong to, and what kind of access they have to computers, such motivation may not even exist or may disappear at varying speeds.

Table 3-1: Different perspectives considering modeling and technology

TECHNOLOGY MODELING	Tutor	Motivation	Reorganization	Citizenship
1) Applied problems in traditional textbooks	Electronic book			
2) Mathematical topic		Create an environment for a topic	Imaginary City (Araújo, 2002)	
3) Projects				Family Project (Skovsmose, 1994)
4) Student and curriculum		Searching on the Internet	Examples in Chapter 6	Examples in Chapter 6

Moving towards the right in the matrix, we can find in the work of Araújo (2002) an example in which the teacher involved in a calculus course for chemical engineering majors applied a combination of modeling approaches 2, 3 and 4 in Table 3.1. He asked the students to choose a theme that had a function associated with it; he specifically asked them to choose a function from their everyday lives. Araújo, who was not the teacher, but rather a participant observer, followed the activities of some of these groups. One of the groups chose an 'imaginary city' as a theme, and the variation of temperature in the city as the specific issue to be investigated. They built a function that would describe the variation of temperature in the city. The students used the computer to build graphs and study algebraically a piecewise-defined function. Araújo argued that the software was a significant part of this collective, also composed of the students and herself as a 'kind of teacher', since they knew she had degrees in mathematics. Due to the design of the software, students were lead away from their original interest and formulated new problems that actually had to do with continuity of piecewise-defined functions and the way that the software *Maple* graphed them. This example in Araújo's (2002) research is interesting for many reasons. The main one is that it shows that any classification of modeling can actually be challenged, as one may see traces of perspectives 2, 3 and 4 in the teacher's proposal. It also points out that sometimes the classification of a given example in a given technology column may be made by the researcher who is analyzing an example developed by a teacher, but it may be reclassified in different cells after the researcher has understood the meaning of the task for the students.

Although the classification has the problems we have pointed out, it can serve as a tool to see trends and possible resonance between views of information and communication technology on the one hand and modeling on the other. A clear case is the example reported in an earlier book of the Mathematics Education Library (Skovsmose, 1994) in which a spreadsheet software is used in the 'Family example'. This approach to modeling is a clear case of the third perspective in which the project, which posed a problem situation, was designed by the teacher and the researcher. In this task, the students have to analyze different situations of families (single mothers, elderly, parents with three kids, etc.). The students from this school in Denmark, who were between 14 and 15 years old, had to decide how to distribute among the families a certain amount of money that the city government had available. There is hardly any discussion regarding the use of technology in this example, but from discussion in other parts of the book and other papers by the same author, it can be inferred that he would support the use of computers in schools based on the citizenship argument. Therefore, even though its classification in the technology columns is not as clear as the type of modeling, we classified the use of the spreadsheets in this category, and we understand that the use of the spreadsheet software with that task allowed the kids to manipulate the data, generate conjectures, assess the decisions, review distributions and so on.

Finally, as will be shown in Chapter 6, we have collected many examples in which columns three and four, "reorganization" and "citizenship" respectively, are linked to the fourth perspective of modeling in which there is an emphasis on the choice of the problem by students. This discussion will be re-visited as we attempt to shed light on them using examples from research by members of our research group, GPIMEM.

6. MODELING AND ITS LIMITATIONS

In order to close this chapter, what is important to notice is that modeling has many different conceptions in different countries and even within the same country, especially if we consider countries like Denmark or Brazil where this pedagogical approach seems to be more popular. It is also relevant to point out that there are different possible ways of seeing modeling and the use of computers, and that such synergy can change the nature of this pedagogical perspective quite substantially. We will return to this discussion when we present more examples of the modeling approach we have been developing and its interfaces with technology.

The diversity of perspectives is even greater if we consider our discussion regarding problem solving and problem posing. We hope it is clear to the reader that the main differences between these perspectives have to do with the internalistic aspect of mathematics and the role of the students in choosing the problem. But it is also fair to say that some of the differences we see in these other trends, when compared to modeling in general, are still present in this trend. For instance, there are authors who find that modeling is just some kind of appetizer for 'real mathematics' that should be taught, and others who do not see major relevance in having students choosing their problems, even though they work with modeling.

However, one difference seems to be more important, and it has to do with computers. In the literature we analyzed, problem posing and problem solving do not stress the role of computers. Of course, part of it has to do with the very notion that modeling in itself was 'born' in a time in which the computer trend was also beginning. But it is also fair to say that regular calculators never really became incorporated into either problem posing or problem solving, even though they were available.

On the other hand, modeling became more popular in applied mathematics due to the development of computers. As we have analyzed in this chapter, and will in the next one, the whole idea of modeling gained considerable momentum with the development of computers and powerful personal computers. The status of modeling changed within the greater field of mathematics due to technology and its association with collectives that have produced mathematics in the last fifty years, and in particular in the last twenty years. Expressions like simulation, experimentation, and the like, which were very foreign in mathematics departments, now at least have permission to enter the coffee room.

Such a movement found room in mathematics education due to the crisis of relevance (Niss, 1977); to a movement to provide citizenship for all (Freire, 1992), valuing social cultural basis (D'Ambrosio, 1985; Bassanezi, 1994); and to change in the nature of the structure of curricula (Borba, 1990). It is important not to forget these roots at a time when there are waves of a new back-to-basics movement in some countries, now dressed in a 'back-to-the-disciplines' kind of argument. We do not believe that modeling means the end of disciplines or that modeling solves all the problems, however. We do believe that modeling, as far as we can see now, has its limitations, and its interaction with technology is just one possible pedagogical approach which is in resonance with it. There are times in which one may want to teach polynomium, or a chain rule, or some kind of technique of solving differential equations in which modeling can become, at a minimum, artificial. We do believe that the study of such topics is different if it is born in a 'modeling atmosphere'. We do not want to claim that modeling is just one more pedagogical approach within a landscape of others. We believe that it should shape curricula, but that other, more structured proposals should also be developed, especially because we believe that some of them are in resonance with the computer medium, in the sense that they explore its possibilities. When we discussed technology in Chapter 2, we stated that information and communication technology did not mean the end of other media such as writing and orality. We want to propose a similar case with modeling: it will not suppress other practices in the classroom, but it will change their nature, and it should co-exist with them. In the following chapters, we will analyze how issues regarding experimentation and visualization have also been transforming mathematics education. We will also show how experimental pedagogical approaches are in resonance with our view of information and communication technology.

Chapter 4

EXPERIMENTAL-WITH-TECHNOLOGY APPROACH: RESONANCE WITH MODELING AND MULTIPLE REPRESENTATIONS

1.   INTRODUCTION

In the last chapter, a perspective of modeling as a pedagogy was discussed. In particular, it was shown that, although modeling has been a very promising perspective in mathematics education, it is not the only perspective we propose to be developed in the classroom. In the different approaches discussed within the Brazilian or the Danish perspective, it was pointed out that it is often necessary to take a detour from a given investigation in order to learn a specific topic in mathematics. In this sense, students may need to step back and learn certain things about functions, analytic geometry, or differential equations so that they can keep investigating the problem they are working on.

A parallel to this pedagogical approach can be found in the activity of researchers. Let's say a research group is investigating how students learn when using a given software. They chose this topic and are very interested in it, but at a certain point, they decide to stop and study how, at a policy level, computers have been introduced into schools in the country where they are developing their study, so that they can place their findings in a broader context. Many of us who develop research do similar things with epistemology, sociology, psychology and other areas in order to inform our studies with new perspectives. Usually on this side trips, we take paths such as the following: individual reading, research group seminars, or a graduate-level course. At least these have been our experiences when we need to stop the main course of our research, at any point, in order to study some topic (which, of course, is also part of the research).

In this example, we have a researcher working on a project. Similarly, students are engaged in their projects when the modeling approach is being used. We just described what researchers do when they come to a stumbling block; but what should we, as teachers, propose that students do when they encounter an obstacle similar to the one we often encounter while doing research? Of course, there are many answers to this question. Sometimes what happens, according to some references, is that traditional teaching with chalk and blackboard is what takes place. Although we believe that traditional teaching has its place and sometimes is the best way to inspire learning, we can think of other ways of doing such a task. Reading is one; in this sense, the student would learn how to read a textbook and learn a given topic with help from other students and the teacher. Forming discussion groups is another, particularly with the opportunities that the Internet provides for collective virtual discussions. Another possibility is the approach that we developed in one of the GPIMEM research projects and labeled 'experimental-with-technology'.

Let us give an example. What should we do when students come to a situation like this while engaged in activities within a modeling project? In one of our Applied Mathematics courses for biology majors, one group of students was studying the influence of different substrates on the germination and growth of a certain plant, considering the size of the seedlings and the number of seeds that germinate in each substrate vs. time. For each substrate, a graph of the number of germinated seeds vs. time was done, and it became clear to the students that they needed to know more about logarithmic function, which seemed to be the prototypic function that fit the data. In the process of studying aspects of the graphs, transformations of log function were often the first steps recommended by the teacher to study the properties of the logarithms involved in the students' projects. Based on activities suggested by the teacher, this group of students then studied the relation between coefficients and graphs of functions $y(x) = a.\log(x + b) + c$ using a graphing calculator. They had already done this with quadratic functions $y(x) = ax^2 + bx + c$ as a means of studying the transformations in the graph when variations in coefficients a, b and c are made. So at this point they employed the experimental-with-technology approach with a different family of functions. In such an approach, students will guess, explore, make more educated guesses, build conjectures, and discuss their findings with colleagues and the teacher. For the time being, their theme of investigation shifted from plants to log function.

In this chapter, we will present this experimental-with-technology approach to the reader. We will discuss how experimentation is seen in mathematics. Later we will present a discussion of how it is seen within the field of mathematics education, presenting how experimentation has been changing the nature of the knowledge produced in mathematics classrooms. Coordination of the experiments done with tables, graphs and algebraic representations has become a relevant issue in the literature that considers the expression 'multiple representations' to be an icon. Finally, we will present how we see the articulation of the modeling approach, multiple representations, experimentation and technology.

2.   EXPERIMENTATION IN MATHEMATICS

Perhaps the best way to start a discussion about the role of experimentation in mathematics is to answer the question: what is an experiment? Webster's Dictionary (1966) shows various meanings for experiment:

1) a test or trial.

2) an act or operation carried out under conditions determined by the experimenters (as in a laboratory) in order to discover some unknown principle or effect or to test, establish, or illustrate some suggested or known truth. (p. 800)

In a newer edition of Webster's Dictionary (1989), we found the following definition:

an operation carried out under determined conditions to discover, verify or illustrate a theory, hypothesis or fact ‖ a method or procedure adopted without knowing just how it will work. (p. 333)

According to these definitions, an experiment is carried out to discover something unknown, to verify the truth of a hypothesis in order to accept or reject it or to provide examples (illustrate) of a known truth, all actions that neither mathematicians nor students could say they have never done. However, there is a contrast between experiment, on the one hand, and the notions of deduction, demonstration and logic, on the other, which are usually associated with mathematics.

An on-line dictionary of philosophy shows the following meaning for the word experiment:

A trial or test of a scientific hypothesis or generalization by manipulation of environmental factors to observe whether what results agrees, or disagrees, with what the hypothesis predicts. (FOLDOP, 2002)

and a more complete definition is presented in a dictionary of education (Good, 1959):

The trial of a planned procedure accompanied by control of conditions and/or controlled variation of conditions together with observation of results for the purpose of discovering relationships and evaluating the reasonableness of a given hypothesis. (p. 215)

From the two previous definitions it can be noticed that, in an experiment, conditions or factors are manipulated and facts are observed in order to prove or disprove a given hypothesis, activities that are related to the physical and natural sciences. In such a perspective, the experimental is associated with the ideas of *founded on, derived from,* or *discovered by* experiment. And recently, there has been a movement, even inside the mathematics community, to value experimentation as an important process in the production of mathematics.

Mathematics is seen for the most part as a science which is formal, deductive, detached from experience and a paradigm of a priori[14] knowledge (Shapiro, S., apud Bicudo, 2002). Although this paradigm has been increasingly challenged, even among those who support this perspective, and despite books such as Hersh (1997) and Davis and Hersh (1985) which state that mathematics is, at a minimum, more than deduction, the main social representation of mathematics continues to be one associated with precision, deduction and logic, with formal proof being considered paradigmatic for mathematics. It would appear that experimentation, in particular, is seen as something which is opposed to such concepts.

According to Schoenfeld (1994), some philosophers and mathematicians have attempted to reconceptualize and redescribe the mathematical enterprise, considering as a main theme that "the *doing* of mathematics is a (somewhat) empirical endeavor" (p.54). In the preface to the first printing of *How to solve it?*, Polya (1945) said:

... mathematics has two faces; it is the rigorous science of Euclid but it is also something else. Mathematics presented in the Euclidean way appears as a systematic, deductive science; but mathematics in the making appears as an experimental, inductive science. Both aspects are as old as the science of mathematics itself. But the second aspect is new in one respect; mathematics "in status nascendi," in the process of being

---

[14] A priori knowledge in the sense of prior to and independent of experience.

invented, has never before been presented in quite this manner to the student, or to the teacher himself, or to the general public. (p. vii)

Polya is shedding light on a frequently hidden part of mathematics, the one associated with the process of mathematical discovery. This process has also been described by Imre Lakatos, who, drawing on the work of Polya as one of his ideological sources, asserted that mathematics is quasi-empirical (Lakatos, 1978). In *Proofs and refutations*, Lakatos (1976) challenged the formalist bastion and mathematical dogmatism, pummeling the certainty and immutability of mathematics:

...informal, quasi-empirical, mathematics does not grow through a monotonous increase of the number of indubitably established theorems but through the incessant improvement of guesses by speculation and criticism, by the logic of proofs and refutations. (Lakatos, 1976, p. 5)

In resonance with this excerpt, we would like to cite some more recent assertions made by Schoenfeld (1994), referring to the nature of mathematical activity:

The *result* of mathematical thinking may be a pristine gem, presented in elegant clarity as a polished product (e.g., as a published paper). Yet the path that leads to that product is most often anything but a straightforward chain of logic from premises to conclusions.

... mathematics is a "hands-on", data based enterprise for those who engage in it. ... It has a significant empirical component, one of data and discovery. (p. 58, emphasis in the original)

All three of these authors recognize an empirical character within the mathematical enterprise, but this kind of philosophical reflection is not common among mathematicians, although experimentation is not new in mathematics. Many of the earliest mathematical results were developed by scientists such as Arquimedes, Galileo or Gauss through physical or mental experiments; that is to say, empirically. Experiments were used as methods or heuristics to arrive at a conjecture, and in recent years, their status is changing inside the mathematics community in the sense that, at least some mathematicians are recognizing that it is worthwhile to show the paths they follow to establish a mathematical result.

We can assert without a doubt that technology has a paramount role relative to the use of experiments in mathematics as well as in mathematics education. Experimental mathematics, and the label itself, are associated with a trend in mathematics that is gaining momentum due to computers. Some mathematicians assert that "mathematics has fundamentally affected technology, most notably computers. It is now clear that the converse will also be true" (Borwein, J.; Borwein, P; Corless, R.; Jörgenson, L. and Sinclair, N., 1995). As an example of this fact, we can cite Kenneth Appel and Wolfgang Haken's proof of the Four Colors Theorem in 1976 with a strong computational aid, which provoked mathematical and philosophical reactions related to the established nature of a mathematical proof. Following is a statement made by Jörgenson (1996), the co-founder and research manager of the Centre for Experimental and Constructive Mathematics (Simon Fraser University, British Columbia, Canada), which aims to explore the interplay between the emerging computer technologies and mathematics:

> Mathematics is experiencing a tremendous upsurge of new activity. Like so many other fields, this is primarily due to computers and related technologies. However there is a notable difference: Mathematics is being fundamentally challenged by these new modes of thinking and discourse, some of which threaten to upset long-standing traditions. A good example is experimental mathematics which represents a recent movement towards empirical and heuristic research practices. While similar methodologies have been in effect for centuries in the sciences, there have rarely been opportunities to perform classical experiments in mathematics. Rather, mathematical knowledge has typically been authenticated on the basis of strict adherence to theorem-proof constructions that are (in general) unreproachable. For many, these new possibilities threaten to undermine the rigor which most mathematicians have held dear. Some doomsayers have even predicted the trivialization and subsequent demise of mathematics in the face of such influences. (http://www.cecm.sfu.ca/projects/PhilVisMath/vis96panel.html)

The above quote suggests that one of the main obstacles for experimentation in mathematics is the fear that experimentation can challenge a methodology that seems to be well established. A similar concern was expressed during a seminar[15] regarding the whole of demonstration that took place in Rio Claro, São Paulo, Brazil in February of 2002. The main papers of the seminar were published as a special section in BOLEMA (2002). Those papers, produced by professionals from the field of logic or mathematics, protected the role of demonstration in mathematics, similar to the way that Jörgenson noted among some members of the

---

[15] Seminar: *Como a demonstração é considerada em diversas áreas de conhecimento?* (How is demonstration considered in different fields of knowledge?). February 22-23, 2002. Graduate Program in Mathematics Education. State University of São Paulo. Rio Claro. Brazil.

mathematics community. Members of the mathematics education community expressed, in the same seminar, a more flexible position in which the importance of demonstrations was recognized, in particular for mathematics majors, but argued that other ways of convincing should be used in the mathematics classroom. Comparisons with demonstrations in other fields, as different as Law, were also made in this seminar. But in all cases, even among mathematicians, it was recognized that a demonstration is also negotiated among peers, and is therefore a social construction. Intuition was even accepted by the mathematicians as important, but not experimentation. In spite of oppositions like these, there is a growing movement inside the scientific community advancing in the direction of experimental mathematics. The existence of research centers such as the Centre for Experimental and Constructive Mathematics at Simon Fraser University (Canada) or the Institut für Experimentelle Mathematik (Institute for Experimental Mathematics) at the University of Essen, (Germany) or the scientific journal *Experimental Mathematics*, edited since 1992 and devoted to experimental aspects of mathematical research, are evidence of such a trend.

At the end of 1995, *The Workshop on Organic Mathematics* was held at the Harbour Center campus of Simon Fraser University. It was hosted by the Centre for Experimental and Constructive Mathematics. Among the articles published in the virtual workshop's proceedings, we found an interesting discussion about experimental mathematics presented by Borwein, J.; Borwein, P.; Girgensohn, R. and Parnes, S. (1995). The authors state:

> Experimental Mathematics is that branch of mathematics that concerns itself ultimately with the codification and transmission of insights within the mathematical community through the use of experimental ... exploration of conjectures and more informal beliefs and a careful analysis of the data acquired in this pursuit.
> (http://www.cecm.sfu.ca/projects/OMP/)

We can highlight some key words in the above excerpt: *insights*, *conjectures* and *informal beliefs* that are connected to the very origin of mathematical discoveries itself. In this sense, the editors of the journal *Experimental Mathematics* stress the importance of experiments as a method of mathematical discovery and note that traditional mathematics journals only accept "elegant, well-rounded and rigorous results". They consider the final discovery as important as the path that lead to it:

> While we value the theorem-proof method of exposition, and while we do not depart from the established view that a result can only become part of mathematical knowledge once it is supported by a logical proof,

we consider it anomalous that an important component of the process of mathematical creation is hidden from public discussion. It is to our loss that most of us in the mathematical community are almost always unaware of how new results have been discovered. It is especially deplorable that this knowledge is not made part of the training of graduate students, who are left to find their own way through the wilderness (Epstein, Levy, de la Llave, 1992, p.1)

It is very interesting to note the editors' strong emphasize on the process of creation in mathematics as a complementary part of the deductive activity. In a side note, mathematicians' concern with the education of mathematics graduate students (future mathematicians) is remarkable, a concern which is rare both in the mathematics and the mathematics education community. The editors of *Experimental Mathematics* also assert that some traditional mathematics journals are reluctant to accept experimental results for publishing. In this sense, they envision the journal *Experimental Mathematics*:

> ... as something akin to a journal of experimental science: a forum where experiments can be described, conjectures posed, techniques debated, and standards set. We strongly believe that such a forum will further the healthy development of mathematics. (Epstein, Levy, de la Llave, 1992, p.3)

In the above excerpt, experimental mathematics gains a broader role; instead of portraying itself as a branch of mathematics, it suggests that it has a role to play for mathematics as a whole. They also add, on the homepage of their journal, that 'experiment' should be considered in a broad sense, as it can be carried out with computers, pencil-and-paper, or with other experimental techniques, such us building physical models.

The previous considerations refer to research in mathematics and reveal visions that are compatible with the opinions of some researchers inside mathematics education. We can say that the experimental mathematics movement is within the boundaries of mathematics, and it is a step forward in the direction of recognizing experimentation in the mathematical realm. Nonetheless, it still seems that experimentation is considered, in some sense, to be a threat to rigorous mathematical development, although it has been of great help in the progress of other sciences. In spite of the space that experimental mathematics is gaining in the mathematics domains, it seems to us that experimentation still has a 'non-scientific' role in supporting mathematical results, since its presence is deleted from the final well-founded proof. But what can be said about experimentation in mathematics education?

3. EXPERIMENTATION IN MATHEMATICS EDUCATION

Inside the mathematics education community, there is an interest in experimental aspects of mathematics. As in the case of mathematics, different views regarding the notion of experimentation also co-exist within the mathematics education community. It is not our goal to emphasize these differences in this book, but to stress the notion that, despite the differences within the mathematics education community, experimentation associated with computers has a paramount role in mathematics education. We would also like to claim that this standpoint does not necessarily mean a rejection of traditional mathematical proof, but a broadening of perspectives to be considered in the processes of teaching and learning mathematics.

We would like to approach experimentation in mathematics education taking the following steps: first, we will refer to some literature on the theme; next, we will discuss the interplay between experimentation and media; and finally we will focus in greater detail on the perspective of our research group, GPIMEM.

In 1996, Gary Davis and Keith Jones (1996) from the Centre for Research in Mathematics Education (CRiME - University of Southampton - United Kingdom) led a discussion group on the psychology of experimental mathematics at the 20th annual Conference of the International Group for the Psychology of Mathematics Education, Universitat de Valencia (Spain). CRiME developed a research project aiming to consider the implications of the rise of experimental mathematics for learners and teachers of mathematics. They state that experimental mathematics has gained respectability in recent years, and that computers are partly responsible for this change. They do not see experimental mathematics as antithetical to mathematical proof, but as

> ... the "behind-the-scene" part of mathematics that never appears in the text-books or journals (until very recently) but which is important in getting a handle on mathematical knowledge - in ascertaining what is likely to be true and what is not. (CRiME, 2002)

The researchers from CRiME assert that mathematicians carry out experimental mathematics before the formulation of a conjecture they believe to be true and before the construction of a "water-tight proof". They view experimentation in mathematics education as being an activity that mathematicians perform before establishing the truth of a theorem.

CRiME seemed to be emphasizing that experimentation should be present in mathematics education since it is more and more present in mathematics. We would like to present an argument which stresses another aspect: experimentation should be more present in schools because computers are more available there, and experimentation is in resonance with collectives which involve computer technology. The use of devices like those designed by Nemirovsky (Nemirovsky and Noble, 1997), the use of geometrical software such as *LOGO, Cabri, Sketchpad* or *Geometricks*, the computer algebraic systems such as *Maple, Derive, Mathematica*, the graphing calculator or the so-called microworlds[16] (Noss and Hoyles, 1996; Noss, Healy and Hoyles, 1997) generate environments that can be considered as laboratories where mathematical experiments are performed, considering experimentation in the broad sense we talked about.

For us, experimentation-with-technology means much more than 'pressing keys' on a calculator, graphing calculator or computer. Once the first wave of resistance to the use of calculators in schools subsided around the 1970's or 1980's, another argument against the use of computers emerged based on the assumption that all students do when they use them is press keys, instead of thinking, demonstrating, and so on. We would agree that just pressing keys may not be a very noble learning activity, just as copying within the paper-and-pencil medium and just memorizing within the oral medium may not be very educational activities either.

We would like to argue that when we are memorizing we are not 'just memorizing'; when we are copying we are not 'just copying', and thus, we want to suggest that we may also not be 'just pressing keys'. Lévy (1993) reviews part of the literature about memory to propose that different media are different forms of memory extension, as initially discussed in Chapter 2. In this way, memory is intertwined with myths, as the circular form of myths enhance the capabilities of memory, which in turn become an important agent in keeping cultural tradition in societies where orality is the predominant medium. Humans-with-orality produce knowledge in different ways, and that knowledge is conditioned by this medium; as they repeat certain things to memorize, they are learning, even though they are performing an 'automatic act'.

Similarly, many of us have already copied a demonstration in a completely automatic way without making much sense out of it. However, we may have also copied a demonstration, or even created one, in a way that the copying was much more than an automatic act, even though it did not lose this characteristic. If one considers that memory is extended through paper-and-pencil, the act of copying also becomes part of the genesis of the humans-with-paper-and-pencil system. Copying can then become part of

---

[16] Referring to the word *microworld*, Noss, Healy and Hoyles (1997) assert "the word has come to connote almost any exploratory learning environment which incorporates a computer" (p. 210).

such a system and may provide part of the reflective power that authors like Powell and Ramnauth (1992) and Buerk (1990) attribute to a more creative kind of writing that they use in their teaching. It could also be argued that pressing keys and dragging the computer mouse also transform our memory and pave the way for demonstrations. For instance, authors like Lourenço (2002) propose that the use of software like *Cabri II* be intertwined with the teaching of demonstration for advanced high school students or entry-level university students.

In a similar way, we believe that if all that is taking place is key-pressing, such an activity is not likely to last long. As we will show in the following chapters, which will focus on examples, key-pressing in an experimentation environment may be associated with the generation of conjectures, with the coordination of multiple representations, with 'proofs', and with a new kind of 'trial and error'- characteristics of what we decided to call the experimental-with-technology approach.

In projects of our research group (GPIMEM), carried out over the past ten years, and the work previously developed (Borba, 1993) we have tried different terminology to label an approach that emphasizes experimentation in mathematics education. Studies developed by researchers from GPIMEM have stressed, in different research contexts, the importance of an experimental approach in mathematics education when technology is present. We would like to briefly sketch some of the research developed by members of GPIMEM in which experimentation was used, in a broader sense, due to computer technology and open-ended tasks; and revealing an experimental character of the knowledge generated by humans-with-media thinking collectives.

Borba and Confrey (1996) present a student's mathematical constructions working with transformations of functions in a computer-based multi-representational environment, showing a process where visualization and experimentation were central to making investigations, conjectures and modifications. Villarreal and Borba (1996) wrote about the experimentation of calculus students in a Brazilian state university, in a computational environment using *Derive* software while trying to characterize extreme of functions. Trial and error, conjectures and refutations were elements that characterized students' work. Souza and Borba (1998, 2000) developed teaching experiments with eighth graders from a public school, working with graphing calculators and with a didactical-pedagogical proposal aiming to study quadratic functions with a predominantly visual and 'empirical' approach. The 'empirical' refers to "the possibility of working with tests, by way of trial and error, where the student has the opportunity to elaborate hypotheses, test conjectures, refute them, or arrive at generalizations" (Souza and Borba, 2000, p. 36). The authors stress that the graphing calculator resources favor these aspects.

The research context of many of the other studies in GPIMEM is an Applied Mathematics course for biology majors at the State University of São Paulo, where two pedagogical approaches were used: modeling, and what the teacher of the course called the experimental-with-calculator approach, in which graphing calculators were used as a vehicle for students' experimentation in response to assigned tasks, making conjectures associated with those tasks and testing them. At the same time, the students elaborate a report of their activities, describing the processes they followed to get possible solutions and final results. The studies developed in this environment raised various research issues. Borba (1997b) writes about the reorganization of the classroom and the kind of mathematical debates carried out by the students. Borba and Villarreal (1998) suggest that thinking is reorganized, based on the experimental approach used to introduce the notion of derivative. Borba, Meneghetti and Hermini (1997) and Borba and Bovo (2002) raise questions regarding the relationship between modeling, graphing calculators and interdisciplinarity. These last two studies stress: the importance of the experimental-with-calculator approach to perform function fitting; the analysis of biological situations chosen by students from a mathematical point of view; and the selection of mathematical explanations based on biological restrictions. Some examples from these studies will be shown in Chapter 7 regarding the experimental-with-technology approach.

In the above GPIMEM research with non-math majors, we have seen that the experimental approach can serve other purposes that go beyond automatic key-pressing. In such pedagogical practices, trial and error initially gives way to some form of 'educated trial and error', in which it is possible that ineffable conjectures may be arising. We can distinguish 'educated trial and error' from simple trial and error when the conjectures and guesses are not generated randomly but on the basis of feedback from previous trials, from key-pressing, and from previously-generated mathematical knowledge. In this experimental approach, students with graphing calculators may conjecture and reject propositions based on a combination of experimentation and logical arguments, and they challenge other students' propositions after they have gained confidence regarding some facts and conjectures. Refutations and demonstrations often arise when the teacher systematizes the different investigations that have been carried out by different groups.

Villarreal (1999) documented many instances describing students' actions in an experimental approach where educated trial and error, conjectures and refutations were elements that characterized their work.

These elements are also present in Lakatos' (1976) description of the logic of mathematical discovery. Although Lakatos' work has no pedagogical intention and refers only to the production of mathematical knowledge in research activities, the characteristics of the students' learning processes in an experimental approach, as we will discuss further later on, are similar to those Lakatos described.

More recently, we have started to think about this process as a way of thinking which is neither deduction nor induction but *abduction* (Cunningham, 1998; Shank and Cunningham, 1996), since the trials are very quickly no longer random. Lakatos described the logic of mathematical discovery, and according to Shank and Cunningham (1996), the logic of discovery is another way of characterizing abduction in the classical sense as described by C. S. Peirce. Abductive reasoning entails the study of facts and the search for a theory to explain them. It is the mode of inference dealing with potentiality: possible resemblance; possible evidence; possible rules leading to plausible explanations; possible diagnostic judgments; clues of some more general phenomenon. Shank and Cunningham (1996) associate this kind of reasoning with the learning of informal sciences and learning via the World Wide Web, and we also find some connections with the learning of mathematics when the experimental-with-technology approach is used.

At this point in the discussion, based on ideas from mathematics and mathematics education and the definitions we found in common dictionaries, as well as dictionaries of philosophy and education, we are able to say that an experimental approach in mathematics education implies:

- the use of tentative procedures and educated trials that support the generation of mathematical conjectures;
- the discovery of mathematical results previously unknown to the experimenter;
- the possibility of testing alternative ways of getting a result;
- the chance to propose new experiments;
- a different way of learning mathematics.

We can also say that the experimental approach gains more power with the use of technology and thus, the experimental-with-technology approach provides:

- the possibility of testing a conjecture using a great number of examples and the chance of repeating the experiments, due to quick feedback given by computers;
- the chance of getting different types of representations of a given situation more easily;

- a way of learning mathematics that is resonant with modeling as a pedagogical approach.

Let us say a word about these last three topics related to the experimental-with-technology approach. Quick feedback is given nowadays via 'the new computer orality', as in the case of software for children in which instructions are given and interaction with humans takes place via a set of pre-recorded messages. As the Internet becomes more and more user-friendly, all kinds of feedback associating orality and visual aspects have become increasingly present. The different kinds of feedback are closely connected to the different interfaces that computer technology offers. Visual feedback was, of course, only possible once computers had monitors as a means of interacting with humans.

Software used in computers and graphing calculators use visualization as the main means for feedback: visual feedback of a calculation made by the calculator, by a table displayed in a computer software, or a Cartesian graph exhibited on a computer screen or on a graphing calculator which, for example, describes a movement recorded by a sensor. The coordination of such representations has become increasingly relevant. The next section is devoted to such coordination, while the next chapter will be fully devoted to visualization.

Finally, the association of experimentation, technology and modeling exhibits a natural resonance for us and contributes to a pedagogical approach that will also be in resonance with our democratic concern: the accessibility of mathematics to everyone, not just for future mathematicians.

4. MULTIPLE REPRESENTATIONS AND MEDIA

From the 1980's up to the mid 1990's, there was a trend in mathematics education to work with multiple representations. As a result of greater accessibility to graphing calculators and computers, the use of multiple representations has been discussed intensively, especially for mathematics topics such as functions that seem to lend themselves to such an approach. Multiple representational software has been developed for computers and calculators at a speed which is difficult to keep up with. Authors (e.g. Borba, 1994, Borba and Confrey, 1996) have stressed the importance of such an approach as it facilitates students' coordination of established mathematical representations such as tables, Cartesian graphs and algebraic expressions. Confrey and Smith (1994) in particular, raised the status of such a discussion when they coined the term 'epistemology of multiple representations'. By this, they were implying that multiple representations could actually change the way students and teachers know mathematics.

Until the mid 1990's, for instance at the proceedings of a working group of ICME-8 (*The role of technology in the mathematics classroom*, Borba, Souza, Hudson and Fey, 1997), which took place in Spain, the use of this theoretical construct was quite intense. However, at the end of the 1990's and early in this century, it can be noted that interest in multiple representations as a theme of investigation waned. A look at the proceedings of PMEs 22, 23 and 25 (Oliver and Newstead, 1998; Zaslavsky; 1999, Heuvel-Panhuizen, 2001) shows that, although the expression is still used in areas like geometry, it is hardly used or discussed in the areas of functions, algebra and calculus, for example, where they were once very popular. Of course there are exceptions to the rule such as Tabach (1999), who still uses such a construct to discuss algebra. There are many possible reasons for this shift in focus: one is that maybe most researchers and teachers are using this concept, and therefore it is part of mathematics education that is 'taken for granted'. However, it is very unlikely that teachers worldwide are using concepts associated with multiple representations, but if this were the case, interesting research could be developed to make comparisons with the earlier research in this area, which was mostly developed with one or two students at a time[17]. The second possible reason is that maybe it is just 'out of fashion', and researchers need to be saying new things in order to obtain grants for research. Thirdly, it could be the case that the problems of research have been solved and there is no more research to be developed. It is not our purpose to analyze the reasons for the decreasing number of papers about multiple representations issues in our area, although it seems to be a worthwhile undertaking.

The above analysis can be found in Borba and Scheffer (in press) in which we discuss how a different look at different interfaces could revitalize the discussion about multiple representations. In this videopaper, we show how sensors, such as CBR, linked to a calculator and open-ended tasks, are suitable for a transformation of the multiple representations discussion, as it involves coordination of body movement (some of our most basic experiences) with standard mathematical representations such as tables, graphs and algebra. A more complete discussion about the role of the body can be found in this videopaper. At PME 25 (Pateman, Dougherty and Zilliox, 2003), it can be noticed that topics related to multiple

---

[17] During the final revision of this book, we learned of the existence of the paper by Patterson and Norwood (2004) that addresses how teachers deal with multiple representations. The article supports our point of view that more research is needed about multiple representations.

representations make somewhat of a 'comeback', even if the specific terminology is not adopted. For our purposes, we simply want to argue that consideration of the body is relevant for mathematics education, and that linked with the many different layers of computer technology interfaces now available, it is possible to amplify the notion of epistemology of multiple representation proposed by Confrey and Smith (1994). Monitors intensified visual feedback. Calculators and computer keyboards intensified the use of fingertips, sensors intensified the connection between body experience and academic mathematical representations. We believe that computer technology and its different interfaces are changing the nature of the senses we use to communicate within a humans-with-media unit. If we see our own body-mind as interfaces, we can propose that part of the reorganization of thinking has to do with different combinations of human and computer interfaces, and that each one of these 'actors' constitute each other. In this sense, we will show examples later in this book that will return us to this discussion. For this chapter, however, it should be emphasized that the new possibilities provided by different interfaces have extended the possibilities of experimentation. Experimentation using sensors, with graphic and table feedback, have been explored not only by us, but by many authors, in particular the group at TERC (Technical Education Research Center), Boston, U.S.A. led by Ricardo Nemirovsky. We believe that these possibilities also provide new opportunities for modeling as students can explore themes that used to be costly to explore, either timewise or financially. In this sense, we believe that, although experimentation was presented earlier as a complement to modeling, as discussed in Chapter 3, it is also opening new doors for modeling as a pedagogical proposal, intensifying its resonance with the view of technology we support in this book. As we hope to show the reader in the next chapter, the discussion regarding multiple representation, experimentation, modeling and the notion of humans-with-media will interact with the one regarding visualization, one of the most explored themes in mathematics education in the last fifteen years.

Chapter 5

VISUALIZATION, MATHEMATICS EDUCATION AND COMPUTER ENVIRONMENTS

Visualization seems to be the main means of feedback provided by computers since monitors transformed the nature of computers. However, the discussion about visualization in the mathematics education community is much broader than this, and is sometimes not even associated with computers. In this chapter, we will discuss how the status of visualization has changed in mathematics education. Unlike experimentation, the subject of visualization, associated with computers or not, has generated an immense amount of literature. Although just a few references have been chosen here, we believe they represent the spectrum of work in this area. They exemplify how issues related to visualization have been addressed by different authors, and how the theoretical view presented in this book may shed new light on this discussion.

1. VISUALIZATION: SOME DEFINITIONS

Research about visualization in mathematics education is widespread. Visualization has been considered as a way of reasoning in mathematics research as well as in mathematics learning. Advantages and disadvantages of visual approaches in teaching and learning mathematics have been characterized. Relationships between visualization and mathematical performance, and between visualization and mathematical giftedness have been studied.

The set of definitions associated with visualization is quite broad. Different ways of seeing it appear in the literature, and these differing perspectives have been broadened and explained in greater depth in recent articles. The terminology also varies: spatial ability, imagery, visual image, and, visualization are terms frequently used and defined.

Lohman (apud Clements, 1981, p. 35) declares that "spatial ability may be defined as the ability to generate, retain and manipulate abstract spatial images". Imagery is "... the occurrence of mental activity corresponding to the perception of an object, but when the object is not presented to the sense organ" (Hebb apud Lean and Clements, 1981, p. 267, 268). Presmeg (1986a) defines visual image as a mental scheme representing visual or spatial information. This definition is purposely broad to include different types of images representing models or shapes and also pictures in the mind, depending on the clarity of the images. According to Presmeg (1986b), this definition also allows for the possibility that verbal, numerical or mathematical symbols be spatially arranged to form an image. Presmeg (1986b) was able to identify different types of imagery: *concrete pictorial imagery*: known as *pictures in the mind*, clear and intensive images; *pattern imagery*: relations represented through a visual-spatial scheme; *memory images of formulae*: some people 'see' a formula in their minds, written on a blackboard or a notebook; *kinesthetic imagery*: images including muscular activities; *dynamic imagery*: moving images.

Focusing specifically on the word *visualization*, we find different characterizations. As stated by Ben-Chaim, Lappan and Houang (1989), visualization encompasses the ability to interpret and understand figural information and the ability to conceptualize and translate abstract relationships and nonfigural information into visual terms. In this case, we can distinguish two processes: interpretation of visual information, and the generation of visual images from nonfigural information. This second process is also present in the words of Eisenberg and Dreyfus (1989) when they state: "many concepts and processes in school mathematics can be tied to visual representations, that is, visual models can be built which reflect (a large part of) the underlying mathematical structure" (p. 1). Thus, they see the process of visualization as follows: this mathematical concept can be thought of in terms of this diagram or that graph; that is, visualization is associated with visual representation.

Zimmermann and Cunningham (1991) point out that visualization in mathematics is a process of forming images (mentally, or with paper and pencil, or with the aid of technology) and using them with the aim of obtaining a better mathematical understanding and stimulating the mathematical discovery process.

Gutiérrez (1996) considers visualization in mathematics as "... the kind of reasoning activity based on the use of visual or spatial elements, either mental or physical, performed to solve problems or prove properties" (p. 9). He states that visualization is composed of four main elements: mental images, external representations, visualization processes and visualization abilities.

Zazkis, Dubinsky and Dautermann (1996) present a broad definition of visualization, valid for other contexts outside of mathematics:

> Visualization is an act in which an individual establishes a strong connection between an internal construct and something to which access is gained through the senses. Such a connection can be made in either of two directions. An act of visualization may consist of any mental construction of objects or processes that an individual associates with objects or events perceived by her or him as external. Alternatively, an act of visualization may consist of the construction, on some external medium such as paper, chalkboard or computer screen, of objects or events that the individual identifies with object(s) or process(es) in her or his mind. (p. 441)

Considering the previous definition, Nemirovsky and Noble (1997) remark that it restricts visualization neither to the student's mind nor to an external media, but rather defines visualization as a means of traveling between them. They also remark that, although a distinction is made between what is external (paper, computer, etc.) and what is internal (in the mind), Zazkis, Dubinsky and Dautermann allow that it is the individual who perceives, and not the researcher who defines, those objects as internal or external.

If we analyze and compare the different definitions we have presented, some similarities may be noticed. It seems clear, from Gutiérrez (1996); Zazkis, Dubinsky and Dautermann (1996); Zimmermann and Cunningham (1991); Ben-Chaim, Lappan and Houang (1989), that visualization in mathematics education is considered to be a process that follows a two way path between students' comprehension and external media. On the other hand, Presmeg (1986a, 1986b) and Eisenberg and Dreyfus (1989) emphasize just one direction of that path. According to Presmeg, the process of forming images has its starting point in external environments, whereas for Eisenberg and Dreyfus, external representations are generated from mathematical comprehension. Comparing the definition given by Zimmermann and Cunningham (1991) with the assertions of Eisenberg and Dreyfus (1989), we can notice that, in the first case, visualization is paramount to the mathematical discovery process, whereas in the second one, its role is secondary, since mathematical concepts are considered as preceding a possible visual representation of them, and thus visualization has no relevant function in the construction of mathematical concepts.

It can be said, as Nemirovsky and Noble (1997) do when they analyze Zazkis, Dubinsky and Dautermann's (1996) definition, that internal/external, or inside/outside-the-mind dichotomies also underlie the definitions just discussed. Nemirovsky and Noble object to this dichotomy, claiming that it is limited because it does not allow the possibility of objects, representations, graphs, etc, being neither inside nor outside, or both inside and outside, at the same time. These authors go further in their analyses and introduce the concept: *lived-in space*: "A lived-in space is not 'carried' by the individual, but created in an ongoing process that involves memories, intentions, and the situation at hand" (p. 105). This concept, as proposed by the authors, is intended to overcome such internal/external, body/mind dualisms.

We recognize that the 'lived-in space' concept entails a step that overcomes traditional perspectives of visualization and motivates new positions not tied to an external/internal dichotomy but related to the lived-in-world constituted by our experiences, activities, humans, non-humans , etc. As will be discussed more extensively at the end of this chapter, the position of Nemirovsky and Noble is compatible with the theoretical perspective previously presented in this book. Technologies such as mathematical software and new interfaces are part of the lived-in-world as well as of the thinking collective (Lévy, 1993) and have a paramount role in the visualization process.

We have presented a landscape of definitions associated with visualization. This process is recognized as being important in mathematics as well as in mathematics education, although its status seems to vary. Nowadays, visualization is closely related to media, to computers in particular, and we want to devote the next sections to a review and discussion of some literature about visualization and media in mathematics and mathematics education, setting the stage for the discussion about visualization as seen through the lenses of our theoretical perspective which emphasizes the role of media.

2.   VISUALIZATION AND MEDIA IN MATHEMATICS

Although we are primarily interested in visualization and media in mathematics education, we cannot overlook the influence of statements made by mathematicians regarding educational activities. Thus, we will briefly introduce some issues about visualization and media in mathematics in order to situate the discussion that has developed within the mathematics education community.

According to Dreyfus (1991), there is a growing unconventional movement in the mathematics community that aims to make visual reasoning an acceptable practice in mathematics in combination with algebraic reasoning. He also says that:

> ... according to this movement, visual reasoning is not meant only to support the discovery of new results and of ways of proving them, but should be developed into a fully acceptable and accepted manner of reasoning, including proving mathematical theorems. (p. 40)

Thurston (1995) goes even further than Dreyfus when he claims that vision, spatial sense and kinesthetic sense are paramount for mathematical thinking. The author indicates that it is easier for students to take in information visually or kinesthetically or through their spatial sense, but that they have difficulties in translating an internal spatial understanding into two-dimensional images. According to him, that is why "mathematicians usually have fewer and poorer figures in their papers and books than in their heads" (p.31).

Thurston seems to claim that there is a special kind of difficulty in expressing internal spatial images, as if he were comparing this to other means of expression, such as algebraic, for instance. We infer this, although he does not explicitly state this. On the other hand, it is quite clear that, for him, there is an apparent inside/outside-the-mind dichotomy, and this path leads from the inside out. In this paper, Thurston also mentions oral and written language as important components of mathematical thinking, not just for mathematical communication.

In agreement with Thurston, Devlin (1997) indicates that two features of current mathematical reasoning that are not taken into account by the classic logic model of mathematical reasoning are: the use of diagrams and visual reasoning procedures, and interactivity and dynamic representations. Although mathematicians make extensive use of these elements when they are solving a problem or trying to convince or facilitate the understanding of the truth of a particular statement to their colleagues or students, they are not yet regarded as legitimate proofs publishable in research papers. Thus, diagrams are not allowed as essential parts of a proof.

In spite of this fact, there are researchers and research institutions that investigate and discuss the role and contributions of visual representations in mathematics and, with the aid of computers, develop powerful visual tools to facilitate reasoning (Hanna, 2000). Discussions about the role of visualization in mathematical research are frequently presented in international workshops or conferences that bring together visualization and mathematics. In 1996, a conference panel, entitled *Mathematical Visualization: Standing at the Crossroads*, was held during the IEEE Visualization meeting (IEEE: Institute of Electrical and Electronics Engineers). Among the questions this panel addressed were:

- Can mathematical visualization consistently lead to new research results, or is it mainly suited for communicating previously discovered work to other researchers or students?
- Does visualization simply contribute to the body of knowledge of mathematics, or can it change the very nature of mathematical knowledge?
- What can visualization contribute to mathematics, and how might that affect the nature of mathematical knowledge?

The above questions, raised by the organizers of the panel in an Engineering Conference, suggest that visualization can change the nature of mathematics itself and the paths of research. This was done cautiously, as if they did not want to touch on the 'sacred cow' of formal proof. There are others, however, who seemed to be prepared to take further steps. Jörgenson (1996), co-founder and research manager of the Centre for Experimental and Constructive Mathematics (Simon Fraser University, Burnaby, British Columbia, Canada), who has participated in numerous inter-disciplinary projects involving visualization, mathematics, philosophy, information and network technologies, states:

> While there is an anecdotal history for the use of mental imagery in mathematics and science, it has no established role save for notable exceptions like geometry, graph theory and most recently the study of chaos and nonlinear systems. Speculatively, this is due to the assumed nature of perception; that it is a largely subjective mode of expression and understanding which is difficult to separate from its "human failings". Typically an insight arrived at in graphical fashion must be transformed into a corresponding analytical result before it can be accepted into the common body of mathematical knowledge. However it may be the case that this is not always possible or even desirable. Some might even suggest that something valuable is irretrievably lost in such a translation.
> (http://www.cecm.sfu.ca/projects/PhilVisMath/vis96panel.html)

Jörgenson recognizes the role of "graphical fashion" in mathematical insight, and he claims that translation into analytical language may take away part of what was discovered. In the language we have been using in this book, algebraic representation can be seen as less important than graphical representation. Such a discussion seems to be in line with the work of some ethnomathematicians, such as Barton (2002), who claims that languages of different indigenous people may be capable of expressing mathematics that is different from the widely-used, academic 'mathematical language'. Analogously, visualization can be seen as a 'language' which can express mathematics that may not be expressible with standard algebraic language.

Opinions like Hanson's (1996), whose research interests include scientific visualization with applications in mathematics, among other areas, can also be found. He states:

> Visualization in general embodies a transformation between a body of knowledge and a picture, or perhaps an interactive animation, capable of representing features of the data to the viewer. In general, the hope is that the displayed features will stimulate associations in the mind of the user that will lead to further insights, suggest new hypotheses to test, and thus advance the progress of science more rapidly than without this methodology.
> (http://www.cecm.sfu.ca/projects/PhilVisMath/vis96panel.html)

According to him, visualization is a methodology to reach new conjectures and to assist the development of science. Here we can recognize certain similarities with Ben-Chaim, Lappan and Houang's (1989) characterization of visualization in mathematics education, when they refer to it as the ability to translate abstract relationships into visual terms. The presence of the inside/outside-the-mind dichotomy is also noticeable. Hanson (1996) also wonders whether visualization can assist mathematical research. In this sense, he finally asserts:

> Subject to continuing debate, however, is the question of whether such techniques will in fact directly contribute to numerous new insights to 21st century mathematics, or whether, for the most part, mathematical visualization will be principally of pedagogical value.
> (http://www.cecm.sfu.ca/projects/PhilVisMath/vis96panel.html)

From Hanson's words, it seems that, although there are some doubts about the contribution of visualization to mathematics, some pedagogical value is attributed to it. Both experts, Jörgenson and Hanson, implicitly assign a secondary role to visualization in mathematics, and we can recognize some similarities between their descriptions of mathematical activities, where visual resources are used, and those described by Devlin (1997).

At this point, we can say that, although many authors stress the importance of visualization, there is an underlying idea that visualization is just a means to attain 'the superior stage of abstraction'. Thus, a visual solution of a problem is considered helpful, but it is 'just' a step toward reaching a final analytic or algebraic solution. Exceptions such as Jörgenson's ideas may just confirm the rule, although they raise hope that things can change.

In this sense, Barwise and Etchemendy (1991) assert that a visual representation is still a second-class citizen, in theory as much as in mathematical practice, and they promote a revision of the formalist doctrine that considers diagrams and forms of visual representation to be "unwelcome guests" in rigorous proofs. The authors recover forms of visual representation, not just as heuristic pedagogical tools, but also as genuine elements in mathematical proofs. Recognizing that this is a "heretical claim" challenging mathematical and logical tradition, the authors give examples showing that it is possible to obtain valid proofs with the use of various forms of visual representations, and they claim that mistaken proofs and fallacious inferences could be generated through propositional reasoning.

In agreement with Barwise and Etchemendy (1991), Davis (1993) presents an article stressing the importance of the so-called visual theorems, and emphasizes the visual as legitimate in the mathematical discovery process. He states: "... the elevation of the visual component of mathematics would restore to the word 'theorem' something of its original flavour: the Greek root of the word means to look at" (p. 341).

We have presented a set of authors' positions related to visualization in mathematics. We can refer to its status on two levels: one associated with its use in mathematical formal proof; and another related to its use in other mathematical activities, such as making a conjecture, solving a problem or trying to explain some mathematical results to a colleague or student. In the first case, visual representations are not accepted as part of a formal proof, but as heuristic accompaniments to proof, inspiring a theorem or its proof (Hanna, 2000), and in the second one, it is more like a peripheral resource or even a pedagogical one. Thus, in spite of the movement Dreyfus (1991) foresaw, and that we refer to at the beginning of this section, it seems to us that there is quite a resistance to recognizing the status of visual reasoning in mathematics research.

Dreyfus (1991) also points out that powerful graphical computers have played an important role in the emergence of such a movement. Numerous authors have written about the role of new technologies in mathematical activities.

Devlin (1997), for example, refers to transformations that new technologies have brought to mathematicians' activities:

Over the past decade or so, the professional mathematician has changed from being a person who sits at a desk working with a paper and pencil to a person who spends a lot of time sitting in front of a computer terminal. The paper and pencil are still there, but a lot of the mathematician's activities now involve use of the computer.... This rapid transformation of mode of working has changed the nature of *doing* mathematics in a fundamental way. Mathematics done with the aid of a computer is qualitatively different from mathematics done with paper and pencil alone. The computer does not simply 'assist' the mathematician in doing business as usual; rather, it changes the nature of what is done. (p. 632, italics in original)

Devlin (1997) considers, like Lévy (1993), that the arrival of a new medium like the computer does not supplant an old one, such as paper and pencil, and he believes that the computer can play a significant role in the mathematician's reasoning process. For example, the possibility of seeing the effects of changing a parameter in an equation may contribute to the generation of new conjectures. This kind of use of the computer, in the acquisition and processing of information, may transform mathematical reasoning. But many authors indicate that this trend, when associated with mathematical proof, is still highly polemic inside the mathematical community (Garnica, 2002); in this case the computer is seen as a "stranger in the nest" (Domingues, 2002). Some authors are more radical, such as Mumford[18] (as quoted in Horgan, 1993), who is critical of "the pure mathematical community [who] by and large still regards computers as invaders, despoilers of the sacred ground" (p. 76).

In spite of this rejection of computers in mathematics, there are authors like Francis (1996) who recognize that:

... it is less important to debate whether a serious preoccupation with computers is relevant to contemporary mathematics than to appreciate the fact that the very future of mathematics is predicated on the ubiquity of the computational paradigm. Just as Newtonian mechanics, optics and dynamics permanently moved mathematics away from static Euclidean geometry, so the computer-dominated information revolution will ultimately move mathematics away from the sterile formalism characteristic of the Bourbaki decades, and which still dominates academic mathematics.

On the other hand, he adds:

---

[18] David Mumford was granted the Fields Medal in 1974 for his research in pure mathematics.

... it is absurd to expect computational simulation and computer experimentation, even at a level of "infinite precision", to replace the rigor mathematics has achieved for its methodology over the past two centuries. Rather than change the nature of mathematics, the computer will change the content of mathematics.
(http://www.cecm.sfu.ca/projects/PhilVisMath/vis96panel.html)

This mathematician shows a particular position regarding the changes the computer may introduce inside mathematics, although he believes those changes will affect not its nature, but rather its contents. His remarks are relevant for our argument. Arguments like Francis' can lend support to the trend to change the contents of school mathematics as well. If we consider his argument together with Devlin's who, like us, supports the idea that computers not only 'assist' mathematicians but transform the nature of what is done, we may be finding more support for changes like those proposed in this book regarding modifications in content and pedagogy due to the participation of computers as important actors.

The previous references are just a sample of different positions on the use of new technologies inside the mathematical community. There is obviously fear of attributing a relevant role to computers in mathematics, but on the other hand, that fear is not insignificant, since we are talking about the nature of mathematics itself. It is a philosophical question that affects the identity of a discipline itself.

From the theoretical perspective presented in this book, however, the computer is seen as part of different thinking collectives. Humans-with-media is a construct that emphasizes 'humans' and 'media' within a given collective. The assertions of some of the authors we have analyzed coincide with our own, in the sense that they focus on the role of paper and pencil or computers in doing mathematics more or less explicitly. In the network of ideas that we presented, we also found that some authors (Francis, 1996; Hanson, 1996) refer to educational issues related to visualization and computers. Let us now turn our attention towards visualization and media in mathematics education.

3.   VISUALIZATION AND MEDIA IN MATHEMATICS EDUCATION

We can state that there exists a 'theoretical' agreement about the pedagogical value of visualization in mathematics teaching and learning. Dreyfus (1991) states: "Visualization is generally considered helpful in supporting intuition and concept formation in mathematics learning" (p. 33), and Bishop (1989) asserts: "There is evidence that there is value in emphasizing visual representation in **all** aspects of the mathematics classroom" (p. 14, emphasis in original).

Visualization in mathematics education has been investigated in the last two decades from diverse perspectives. In 1989, the journal *Focus on Learning Problems in Mathematics* published numbers 1 and 2 of Volume 11 with the title *Visualization and Mathematics Education*. The guest editors, Theodore Eisenberg and Tommy Dreyfus, present the aims of the volume:

> ... to emphasize some of the positive effects of visualizing in mathematical concept formation and to show how visualization can be used to achieve more than just a basic, procedural and mechanical understanding of mathematical concepts. (p. 2-3)

They also state that many processes and mathematical concepts are related to visual interpretations, and that visual models can be constructed that reveal a large part of the underlying mathematical structure. For them, the main point is to raise issues regarding the pedagogical and didactical power of visualizations, and problems that may arise from that visual modeling process in mathematics education. According to these authors, this discussion has become more relevant because of the increasing use of computers in mathematics classrooms, the improvement of visual representation, and the possibility of students modifying that visual representation. Thus, the authors allude to the richness and complexity of the multirepresentational introduction of a concept.

In the same volume, Bishop (1989) presents a review of research on visualization in mathematics education. He states that, for the last 100 years, mathematics educators have been interested in visual and figural representation of mathematical ideas in the work of individuals as well as in the process of teaching such ideas. He refers to the power of "visual aids" to introduce complex abstractions in mathematics. He also recognizes that the computer is broadening the possibilities of visualization, and that its presence in the mathematics classroom stimulates a great deal of research and development in the area. Although more than 10 years have past since Bishop published his paper, his statements are still up-to-date.

Two years later, after this special volume of *Focus on Learning Problems in Mathematics*, the Committee on Computers in Mathematics Education of the Mathematical Association of America published a volume entitled *Visualization in Teaching and Learning Mathematics*. In this volume, it is suggested that there has been a renaissance of interest in visualization, mainly due to technological developments and their possibilities in many scientific areas. Visualization is considered as a tool for mathematical comprehension. Zimmermann and Cunningham (1991) indicate that, to reach that comprehension, it is necessary to take into account that visualization, which is usually associated with graphical representations, occurs not as an isolated topic, but inside a mathematical context that also includes numerical and symbolic representations.

Even though the introduction of computers in mathematical teaching and learning contexts assigns a new role to visualization in mathematics education, the visual/symbolic dichotomy and the supremacy of rigor discussed in the previous section still persist in many publications. Authors like Tall and Dubinsky and their collaborators, concerned with advanced mathematical thinking (enrolled since 1985 in the PME Working Group with the same name), have been involved in research considering, among other topics, symbolic and visual aspects of mathematics. We will discuss some papers to introduce their perspectives.

Tall and Thomas (1989), for instance, state that the mental activities usually valued more highly are symbolic and logical, and that the visual and holistic ones are less stressed. The authors discuss the use of computers to encourage a more versatile teaching approach, including both types of mental activities, since the traditional lecture approach leads to a narrow symbolic interpretation, and the use of computers provides a visual framework "...for the mental manipulation of higher order concepts" (p. 117). At the same time, however, this visual framework that supports the development of higher order concepts is also considered to be an environment where algorithmic processes (associated with a sequence of computational commands) may arise. In this respect, Monagham, Sun and Tall (1994), in a study of students working with the concept of limit in a computational environment, point out that the computer shows a product (the result of a particular limit) but hides the process, since the students only know the sequence of orders given to the computer through the software commands. Thus, they warn about the possibility of students learning only the algorithmic or mechanical processes associated with such commands in computational environments.

Dubinsky and Tall (1991) cite another limitation of the use of computers. They state that the software that perform symbolic manipulations (*Mathematica, Maple, Derive*, etc.) are powerful tools, but they warn it is misleading to believe the computer provides an easy way to acquire mathematical knowledge. The computer's rapid execution of mathematical algorithms does not guarantee the understanding of the concepts. In order to acquire such understanding, the authors assert that it is necessary to elaborate appropriate educational proposals to guide the learning and teaching processes. They state that visualization and symbolic manipulations have to complement each other in order to contribute to a deeper mathematical understanding.

Related to the teaching of advanced mathematical concepts, Tall (1993) alerts us to the need to be attentive to the formal definitions and deductions as well as to the complex mental images associated with them, when certain examples are used to explain certain concepts. Thus, conceptions arising from those images that do not verify the formal theory may become potential obstacles to mathematical understanding. Nevertheless, the author recommends the use of such images so that:

> Instead of allowing the experiences to **implicitly** coerce the individual into conceiving properties which conflict with the theory, the strategy is to use **explicit** -even flawed- imagery to stimulate the imagination. (p. 244, emphasis in original)

In this way, Tall asserts that the computer is a rich source of visual and computational images that makes the exploration of mathematical concepts possible. Tall (1991) presents some details about these ideas, stressing the importance of visualization in calculus, denoting strengths and weaknesses of visual images and proposing a graphical approach to calculus through the computer. He indicates that:

> ...to deny visualization is to deny the roots of many of our most profound mathematical ideas. In the early stages of development of the theory of functions, limits, continuity and the like, visualization was a fundamental source of ideas. To deny these ideas to students is to cut them off from the historical roots of the subject. (p. 105)

In this sense, the function of the software is paramount, providing the students with the opportunity to explore mathematical ideas, analyze examples and counter-examples, and then gain the necessary visual intuitions to attain powerful formal insights. However, it seems to us that, although visualization is recognized as relevant, the final objective continue to be the rigorous mathematical proof, as we already reviewed within communities of mathematicians.

In ICME 8, Tall (1996) proffered a conference talking about the complementary function of the visual and the symbolic in mathematics. He points out that concentrating on the symbols may lead to an approach that favors the memorization of procedures that become more complex as the number of rules increases. On the other hand, exclusive concentration on the visual may give insights into what happens in restricted contexts with a limited power of generalization. Thus, in this sense, the computer may perform complex algorithmic tasks, but it may also generate an environment that makes it possible to relate the visual and the symbolic.

The perspectives of researchers associated with the Advanced Mathematical Thinking Group show that, although visualization is considered a fundamental process in mathematical learning, and computational environments have a relevant role in that process, some authors suggest the need to subordinate the visual to the symbolic because of the limitations of the visual approach; however, it must be acknowledged that uncontrollable images, rules and improper understandings exist within mathematical formalism as well. We can also infer that these authors regard the computer as a tool that expands human memory, increases the velocity of feedback, and enhances the possibility of generating images, that would otherwise be inaccessible, but they do not consider the potential role of computers in the reorganization of thinking and the changes in contents or teaching strategies. The following quote illustrates such a traditional position regarding the use of mathematical software in mathematical teaching:

> A symbolic manipulator is a *tool* - a very powerful tool - but any tool can only be used to its fullest capabilities by those who know how to use it. The situation is parallel to the use of simple calculator: they do not teach a child how to add (or divide), but they are useful tools for adding or dividing when one knows what arithmetic is all about. Once one knows how to cope with small numbers, perhaps the calculator can be used to investigate facts with much larger numbers. Likewise, symbolic manipulators are likely to prove more useful - as they have proved useful in mathematical research - once the students have progressed to the stage of knowing what the tool is being used for. (Dubinsky and Tall, 1991, p. 236)

According to Villarreal (1999), this position has no support within the concepts of the thinking collective (Lévy, 1993) or the theory of reorganization (Tikhomirov, 1981). In the above quote, there is not even the suggestion that mathematics could be learned *with* the computer. Making a simple analogy, we could say that it is necessary to learn mathematical contents before we could read a mathematical book or work with paper and pencil. Although it seems an absurd analogy, it sheds light on a position that is popular in mathematics education. Books, paper and pencil are media that allow mathematical learning and comprehension, but they are so incorporated into school activities that their influences on the construction of mathematical knowledge are almost imperceptible or invisible. And this brings us to the main thesis of this book: knowledge is always produced by collectives of humans-with-media.

The effects of predominant media (orality and writing) also appear in a computational environment. Observing the activities of students working in a computational environment, Villarreal (2000) identified two different styles of thinking and approaches to dealing with mathematical questions: an algebraic approach and a visual approach.

An *algebraic approach* in the process of mathematical thinking would be characterized by:

- Preference for algebraic solutions when graphical solutions are also possible.
- Difficulty in establishing graphical interpretations of algebraic solutions.
- The need to run through the algebraic, when a graphical solution is requested.
- Facility to formulate conjectures and refutations or generate explanations based on formulas or equations.

In this case, the computer is not used very much, and the calculations that could be carried out on the computer are done with paper and pencil or just mentally. Rules are remembered to justify diverse mathematical issues. These are the traditional activities performed in a standard mathematics classroom.

A *visual approach* in the mathematical thinking process would be characterized by:

- Use of graphical information to solve mathematical questions that could also be approached algebraically.
- Difficulty in establishing algebraic interpretations of graphical solutions.
- No need to first run through the algebra, when graphical solutions are requested.
- Facility in formulating conjectures and refutations or giving explanations using graphical information.

In this case, the computer is used to verify conjectures, to calculate, and to decide questions that have visual information as a starting point.

Although the characteristics of each approach are presented separately, this does not mean that algebraic and visual approaches are exclusive or disjointed in mathematical activities. The same person can work with an algebraic approach or a visual one, depending on the problem and the media she/he is interacting with. As other authors have concluded, algebraic and visual representations necessarily complement each other in the process of mathematical learning.

As we have discussed in the previous chapter, the construct of multiple representations gained power due to the different feedback that computer interfaces provided to mathematics students. Graphical representation and the discussion about visualization gained a life of their own, as we have seen in this chapter, even though they are still connected to the notion of multiple representations. Borba and Confrey (1996) refer to the possibility of coordinating multiple representations (graphical, numerical, algebraic) in a computational environment. Assigning a relevant status to visualization in mathematics education, they assert that visual reasoning is an empowering form of cognition that implies the need to give the students time, opportunity and resources to elaborate constructions, investigations, conjectures and modifications. The authors also state that visual mathematics supported by the use of computers constitute a model to attract those students who, explicitly or implicitly, reject the hegemony of algebra.

Borba (1993, 1995b) points out that traditional media used in the mathematical realm, paper and pencil, favor the algebraic approach to mathematical questions, whereas computational media encourage approaches where visualization has a paramount role. He indicates that: "In mathematics education, there has been a tradition in the teaching and learning models that emphasizes knowing a given phenomenon primordially through algebra" (Borba, 1995b, p. 72).

He then goes on to suggest that the computer may come to enable a breakthrough in the hegemony of algebra, assigning more value to visualization, and providing the students the possibility of visual approaches to learn mathematics.

Students' preferences regarding the use of visual approaches to mathematics have also been studied by several other authors. The research carried out by Eisenberg and Dreyfus (1991) is revealing. They observed that many students are reluctant to accept visualization and prefer an algorithmic treatment of the problems. The authors analyzed three reasons for this reluctance: a cognitive one (visual thinking makes higher cognitive demands than algorithmic thinking); a sociological one (it is natural that students choose analytic rather than visual procedures since school mathematics is usually linearized and algorithmitized); and one associated with beliefs about the nature of mathematics (mathematics is nonvisual).

In another paper, these same authors (Eisenberg and Dreyfus, 1989) indicate that visualization has a limited role in mathematics curricula, although it is emphasized as being of great importance in the literature. As evidence, they mention that skillful students, as well as mathematical researchers, do not have a tendency to see mathematical concepts in a visual way. This reluctance may have its roots in the way mathematics is presented and communicated by teachers and researchers: in oral or written form.

A recent study from Stylianou (2001) reveals a change regarding the use of visual representations in written solutions of mathematics problems performed by advanced mathematics undergraduates in the United States. According to the author, the students participating in the study were willing to use visual representations, although they had "little training associated with this skill" (p. 232). The author compares the results of his study with

Eisenberg and Dreyfus' research, explaining that the change in students' attitudes towards visualization is due to the fact that visual representations are now part of their mathematics curricula since the calculus reform, initiated in the mid-1980's, and has gained widespread acceptance. But we also find studies, such as Aspinwall, Shaw and Presmeg (1997) warning that "uncontrollable images" can constitute barriers to the construction of mathematical meanings in calculus.

Previous investigations have studied the relations between visualization and the mathematical performance of students who have visual preferences to process mathematical information (Presmeg, 1986a, 1986b). These studies classified students into 'visualizers' and 'non-visualizers' and claimed that the latter have better mathematical performance than the former (Presmeg, 1986a). Presmeg's work made important contributions in the area of studies about visualization, but it is essential to point out that, in those studies where visualization and mathematical giftedness were related, the students considered to be high-achivers, who were generally non-visualizers, were evaluated using tests from mathematics curricula that favor non-visual styles of thinking. It is not recognized that the advantage of non-visualizers over visualizers is almost natural, considering that environments of traditional learning do not usually encourage nor favor the development of visual strategies to approach or solve mathematical problems.

In Presmeg's (1986a, 1986b) studies, we can note the influence of curricular issues associated with visualization; i.e. whether or not visualization is a valued part of the curriculum. Some recent papers have made note of the influence of curricula and teachers' beliefs in the development and acceptance of visualization in mathematics education. Cruz, Presmeg and Güemes (2001) present a case study where a teacher gave little recognition to the successful solution of a problem coming from an eighth grade student who used rich imagery and creative methods based on visual insights. The authors explain that the curriculum, established by the educational institution, is generally based on control, and thus teachers attribute value to the concepts and methods they explain in class because they can control them. Thus, since visual methods were not taught, the teacher did not consider its value as a mathematical resource. Habre's (2001) work presents an experience from a university-level Multivariable Calculus course developed in a computer environment encouraging teaching strategies that emphasizes visualization. The study shows that there are students who avoid visual reasoning, even though it is paramount to the understanding of certain topics of calculus. The author states that it is difficult for students coming from traditional mathematical instruction to overcome rooted attitudes and assimilate the idea of thinking visually.

The references we have presented show different positions associated with visualization inside the mathematics education community. There is a clear trend toward recognizing the relevance of the visualization process in mathematics teaching and learning situations. But we can also note an overtone of mistrust regarding visualization that may have its roots in the influence that the scientific practice of mathematics itself has on pedagogical practices. Meanwhile, it is necessary to establish a distinction: although visualization may be considered a second-class citizen in scientific mathematical production, it should not be considered so in mathematics education, because:

- Visualization constitutes an alternative way of accessing mathematical knowledge.
- The comprehension of mathematical concepts requires multiple representations, and visual representation may transform understanding in itself.
- Visualization is part of mathematical activity and a way of solving problems.
- Technology with powerful visual interfaces is present in schools, and its use for teaching and learning mathematics requires comprehension of visual processes.
- If the contents of mathematics itself may change due to computers, as proposed by some mathematicians, it is clear at this point that mathematics in schools will undergo at least some kind of change.
- Although proof is seen as the official route to truth in academic mathematics, it should not necessarily be transposed to the mathematics classroom at all school levels.

Computer technology stresses the visual component of mathematics, changing the status of visualization in mathematics education. This is not an irrelevant or minor change, and this is particularly apparent if we consider the main theoretical construct of this book: the notion of humans-with-media. The media used to communicate, represent and produce mathematical ideas conditions the type of mathematics that is made and the kind of thinking to be developed in those processes. At the same time, the visualization process reaches a new dimension if one considers the computational learning environment as a particular thinking collective, where students, teacher/researcher, media and mathematical contents reside together. Within this collective, the media acquire another status. For example, Nemirovsky and Noble (1997) assert that tools such as computers, graphing calculators or physical devices may be thought of as "... 'conversation pieces' rather than devices for helping students to internalize a particular visual representation" (p. 103). In other words, the role of media in the visualization process goes beyond the simple act of showing an image.

On the other hand, the importance of the analytic in mathematics curricula (in spite of the search for a balance), the necessity of exact and single solutions, and the limited value attributed to the visual and experimental in mathematics education could be associated with the predominant technologies used in mathematics classrooms and in the process of mathematical production until very recently: orality and writing. Negative reactions towards educational approaches using computers to study a specific mathematical concept can be found, like Mac Lane's (1996): "Teaching, by text and talk (lectures) to convey ideas has been and will be the medium to convey hard-won ideas to new thinkers" (p. 330).

This kind of reaction is common, especially inside the mathematics community. But, we believe that, if the computer integrates an educational thinking collective, it is necessary to generate educational proposals considering the ways of thinking, organization of knowledge, and the changes in the personal relations inside the classroom that the computer encourages.

4. VISUALIZATION AND HUMANS-WITH-MEDIA

As we review the literature, in mathematics education as well as mathematics, we can see that visualization is predominantly attributed a second-class position, even though there seems to be a consensus regarding its importance. On the other hand, although it is not the dominant point of view, there is a growing and increasingly dissident perspective that attributes heuristic value to visualization. Most of it seems to be due, both in mathematics and mathematics education, to the influence of computers.

Some, like Francis (1996), say that the content of mathematics will change due to computers. Others, such as Devlin (1997), go even further and say that computers are changing mathematics itself. Jörgenson (1996) claims that sometimes it is not even desirable to have visualization as a first stepping stone towards analytic forms of expressions, since some of their content may be lost.

A similar situation is found in mathematics education. For some, visualization is just a first step towards more formal mathematics, which is what is considered relevant. For others, like Borba and Confrey (1996), the road is the opposite, as they argue that an initial visual approach to mathematics may attract those students who reject algebraic approaches. But the secondary role of visualization is prevalent. Similarly to mathematics, the recognition of visualization in mathematics education came before the popularization of computers, but at the same time, it became more important after personal computers and graphing calculators became more popular. There are authors who seem to attribute equal value to the roles of algebraic, tabular and graphical representations, sharing the perspectives of Confrey and Smith (1994), Borba and Confrey (1996) and Villarreal (1999). In this 'epistemology of multiple representations movement', there is a trend to say that multiple representations are valid to construct mathematics, but also to accommodate different students and teachers who may have different ways of knowing to begin with. The production of mathematics is associated with the coordination of graphical representation with tabular and algebraic representations. Computer packages that make such coordination possible are central to such a perspective. Technology is seen as shaping humans, and humans as shaping technology in the so-called 'intershaping relationship'.

In all these perspectives in mathematics education, including that just described, computers and humans are considered as separate, different units. Keith Devlin, from the field of mathematics, is the author who seems to most closely share our view of technology as integrated with humans. According to him, computers do not assist humans in making mathematics; they change the nature of what is done, suggesting that different collectives of humans-with-media will produce different mathematics; for example, the mathematics produced by humans using only paper and pencil will be different from that produced by humans-with-computers.

On the mathematics education front, some like Dubinsky and Tall (1991) suggest that the role of computers is secondary as far as knowledge construction is concerned, since one must know mathematics before using them, and visualization is at most secondary in terms of mathematical knowledge.

We believe that, if we adopt the notion of humans-with-media, we will be distinguishing ourselves from those who attribute a secondary role to different technologies of intelligence as well as those who suggest that visualization is either internal or external. As we consider the humans-with-media unit, we already establish the central role of the medium, since different media like orality, writing and computers reorganize our thinking. In approximating technology and humans, humans may have physical interfaces like skin, but our cognitive boundaries are not well-defined. Nemirovsky and Noble (1997) suggest that our experiences, memories and intention are carried with us. We believe that our view is compatible with the ones they hold, in the sense that we do not see the dichotomy between internal and external visualization seen by almost every author analyzed in this chapter. The experience we are having, or had, with a given media is part of this human-with-media unit, even if it is not available at that very moment (Borba and Villarreal, 1998). It is in this sense that we believe that there is an ongoing process between 'internal' and 'external' representations in which the two are so closely associated that such dichotomies no longer make sense.

Multiple representations are compatible with such a view if we extend this notion to consider graphical representations on paper that are qualitatively different from the ones in a computer package such as *Derive* (see Villarreal 2000), since humans-with-media is seen as a collective that changes as new media or new humans become part of it.

The humans-with-media construct does not allow for the external and internal dichotomy since the boundaries are no longer clear for the cognitive being. In the case of visualization, what we see is always shaped by the technologies of intelligence that form part of a given collective of humans-with-media, and what is seen shapes our cognition. Having analyzed this notion of humans-with-media from a philosophical and historical standpoint in Chapter 2, and discussed pedagogical approaches which are in resonance with this view in Chapters 3 and 4, as well as the notion of visualization as seen from this perspective in the current chapter, we will now present, in the next chapters of the book, a series of research examples that we hope will shed new light on this theoretical discussion.

Chapter 6

MODELING AND MEDIA IN ACTION

1.   INTRODUCTION

In Chapter 3 we showed different perspectives of modeling, viewed as pedagogical strategies, and how they can be contrasted with other well-developed trends in mathematics education, such as problem posing and problem solving. We also discussed different perspectives regarding the way information and communication technology (ICT) interacts with different ways of understanding modeling in educational settings.

In this chapter, we will analyze some examples from our research and how this interaction comes to life in the classroom. As the reader will notice, these examples do not cover most of the cells of the matrix presented in Table 3-1 in Chapter 3, but rather are basically from the last two columns and the bottom row. In other words, they result from a theoretical perspective that envisions reorganization of thinking when different media are used, from a political perspective that sees access to computer technology as an aspect of full citizenship, and a view of modeling that stresses the participation of students in curriculum design through choice of problems to be investigated. As a result of this interaction, this chapter will attempt to show how modeling, viewed as a pedagogical approach, is transformed as different humans-with-media collectives produce knowledge in the classroom.

2.   MODELING IN A MATHEMATICS COURSE FOR BIOLOGY MAJORS

Most examples will come from an eleven-year-old research project in which modeling has been combined with the experimental approach in a first year pre-calculus/calculus course for biology majors. The syllabus of the course, which has been taught by Borba since 1993, includes functions, derivative, integral and applications. The official name of the course would be translated as *Applied Mathematics*. Diverse kinds of activities take place in the classroom, from traditional solving of problems in the book, to others, which are intertwined with the modeling approach. For instance, we developed the experimental-with-technology approach discussed in Chapter 4. In this approach, open-ended problems posed by the teacher are designed to involve the graphing calculator or a plotter. Other activities include writing reports about investigations in this perspective, and reading selected sections of the textbook, which take place in the classroom. The problems the teacher proposes to students vary from: "Explore the changes in graphs of quadratic functions $y = ax^2 + bx + c$ when parameters a, b and c are modified", to "Construct a parabola using straight lines only", which is used to introduce derivative (see Borba and Villarreal, 1998). We mention these activities because we have observed that, in some modeling projects, the students have, on occasion, used strategies they developed while working with graphing calculators.

On the first day of class, the teacher introduces himself and any graduate students that may be observing the class for research purposes, and then explains the way the course is conducted. He distributes a sheet explaining, among other things, that 30% of the grade of the course will be assigned based on a project for which they will choose the theme. There are no restrictions or pre-established criteria regarding the selection of the theme nor its association with mathematics. Students usually react with surprise to this idea, as this is not part of their past educational experience and is unlike most of their other coursework. They want the teacher to offer clues as how to proceed, or ask if such-and-such a theme is acceptable. In response, they are usually told to outline their project plan on paper, and are assured that the teacher is on their side, and will even co-author their paper if they wish. This, of course, does not imply actually helping them write the paper, but rather challenging the students and helping them define the limits of the problem, so that the task is of reasonable size. At the end of the course, students present their modeling projects orally during class hours, with each group having from 20 to 30 minutes to present their project, followed by the same amount of time for questions and debate. The audience for the presentations includes the teacher, the students, usually one or two math majors or math education graduate students who develop research in the classroom, and sometimes students from previous Applied Math classes who have been invited by the teacher or the presenters.

The teacher uses different criteria to evaluate the students' projects. There is, however, one overall guiding principle: students are reminded often that the 'amount of work' is a large component of the group grade, as opposed to the usual criteria of 'right or wrong' that they are accustomed to, particularly in mathematics education. They are also told that the written versions of their projects - including the drafts and the number of drafts presented to the teacher throughout the course - as well as their oral presentations, will be considered. Some groups hand in as many as six versions, varying from a half-page initial version, to the final version, which may be as long as twenty pages. The mathematics employed in the projects is taken into consideration, but in a different way than it usually is, as, for example, on standard tests. Students lose points only if they fail to follow one of the paths of investigation suggested by the teacher or by one of the members of the group. A project that does not have 'a lot of mathematics' can receive a good grade. What counts, overall, is the investigation developed, and not the result.

The themes chosen by the students vary greatly, ranging from those that are closely associated with biology, to others which have nothing to do with it, and a few that most readers might consider 'mathematical'. For instance, one of the groups in 2001 chose to study the 'theory of fractals'; others have chosen 'bees' (1993), 'turtles' (1994), 'plants that disguise to survive' (1996), 'cloning' (1997) and 'who is the father?' (2001), featuring the discussion regarding the paternity of children. Other themes, which are unrelated to biology, or at least not directly related, included: 'abortion' (1997), 'a brief history of music' (1997), 'students' knowledge regarding the food they eat at the cafeteria' (1997), 'reforestation and riparian forests' (1999) and 'global warming' (2000).

Since we have had almost one hundred student projects to date, a team has been necessary to analyze all the data. Every year since 1995, we have usually had a technician, who films the parts of the class related to the modeling and the experimental-with-technology approach, and an undergraduate research assistant, who learns about the perspective developed by the teacher as she/he attends the class and helps in the analysis by doing a summary of each project. In addition, over the years, a number of graduate students have developed their independent projects associated with this educational setting. For the last three years, a masters student, Ana Paula Malheiros (2004) has been analyzing most of the 100 student projects, seeking an answer to her research question regarding the nature of the mathematics developed in the various projects.

In the process of reading the many projects handed in by the students, the teacher is also engaging in research, analyzing the data, which has been naturally constructed in the classroom. After the course has ended, the analysis continues, involving the identification of specific episodes that are of particular interest regarding one research question or another. These episodes are then analyzed more fully, looking more closely at the video clips, project versions, and class observations associated with them.

This is the procedure we follow to investigate the role of technology in the development of their projects. We want to investigate the role of different technologies of intelligence as different human-with-media collectives develop their projects and produce different mathematical knowledge. In particular, since we have this set of data from classes taught from 1993 through 2003, we can observe how different media change the production of knowledge in different ways. In 1993, computers were rare at UNESP, and the few that could be found included one or two Macintosh's with scarcely any software, and IBM-compatible PC's with some software but little power in terms of processing, and with interfaces, like DOS, that were very user-unfriendly. UNESP has not been a leader in equipping its laboratories and classrooms with computer technology in Brazil, nor has it been among those in last place. In ten years, however, things have changed dramatically at this university, and in particular, in the classes taught by Borba. First, a few graphing calculators were acquired. By 1995, there were enough available that they could be used in the classroom for a significant amount of time, and in recent years, they have even been used in assessment, although only exploratory research has been developed in this area. Since 1996, GPIMEM has had a research lab equipped with computers, which have been used for different kinds of research developed by various researchers, including the teams that have conducted research in the classes Borba has taught over the last ten years. Of course, since 1996, the Internet has gradually become available on campus, first only for professors, and since 1999, more and more pervasively.

This adoption of new technologies means that the humans-with-media collectives that produce knowledge have been changing over the past ten years, in a sense mirroring some of the historical developments previously discussed in Chapter 2. This same historical process of technological change has also affected the modeling process in different ways, as we will show in this chapter. We will also present examples and, at the same time, show how technologies of intelligence shape the modeling approach and the knowledge that is produced by different collectives of humans-with-media.

3. **MODELING AND HUMANS-WITH-TEXTBOOKS-EXCEL-PAPER-AND-PENCIL COLLECTIVES**

In 1999, a group of five students chose to research chloroplasts, a part of the cell that is responsible for the transpiration of vegetables and for photosynthesis. This part of the cell, which is present in plants, is responsible for the transformation of the sun's energy into chemical energy, the generation of oxygen and the incorporation of CO_2 into the plants. The students analyzed the growth of chloroplasts as a function of time, and, based on their research in biology books, created the graph in Figure 6-1, saying that the equation for it was $f(t) = 2^{(t/20)}$.

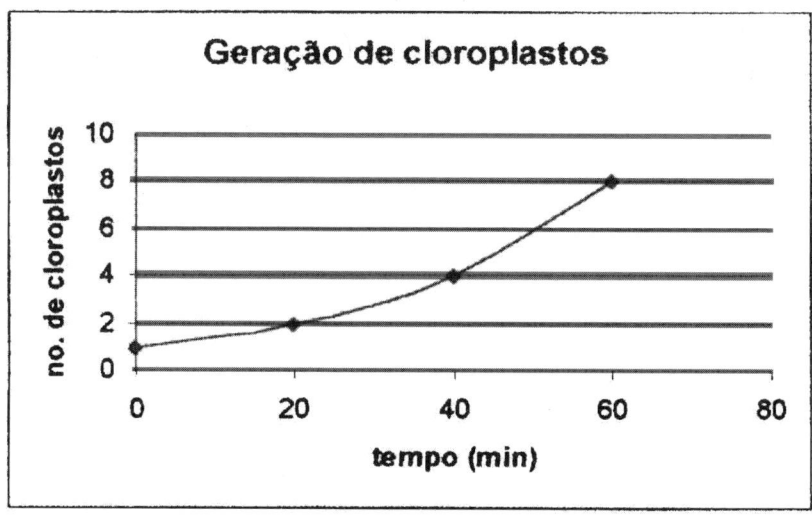

Figure 6-1. The growth of chloroplasts as a function of time. "Geração de cloroplastos", "no. de cloroplastos" and "tempo" mean Chloroplast generation, number of chloroplasts and time, respectively.

The algebraic expression shows the results they found: that the number of chloroplasts doubles every twenty minutes. Of course, this model could have been refined with further investigation so that other factors could have been included in the model, but neither the students nor the teacher chose this path. On the other hand, in the written version of their work, they also pointed out that the derivative of this function was positive and increasing, although they argued that they were unable to evaluate its algebraic expression at this point in their coursework.

The students analyzed the chemical equation that transforms CO_2, H_2O and sunlight into a compound that is rich in energy and O_2. Photosynthesis releases O_2 into the air; respiration of the plant generates CO_2 and water. Based on their review of the biology literature, they found that the rate of photosynthesis is 30 times greater than the respiration rate at its peak. They analyzed this phenomenon, which involves, at a minimum, chemistry, biology and mathematics, and generated several graphs and equations with the *Excel* spreadsheet. At the end of this process, they generated two graphs showing that the respiration rate of the plant is constant as sunlight changes, and that the curve, which represents photosynthesis, has an "s" shape. The teacher informed them that there is a name for that type of curve, logistic, and pointed out the section in the textbook that they could refer to for further study, with the help of the teacher if necessary. They demonstrated in one of the written versions of their project that they had some understanding of the logistic curve, although their graph had neither specific values nor measurement units for sunlight or photosynthesis, and was taken from a biology book in which they were interested only in the quality (shape) of the graph and not its specific values. They did generate the graphs of respiration and photosynthesis in the same frame, as shown in Figure 6-2.

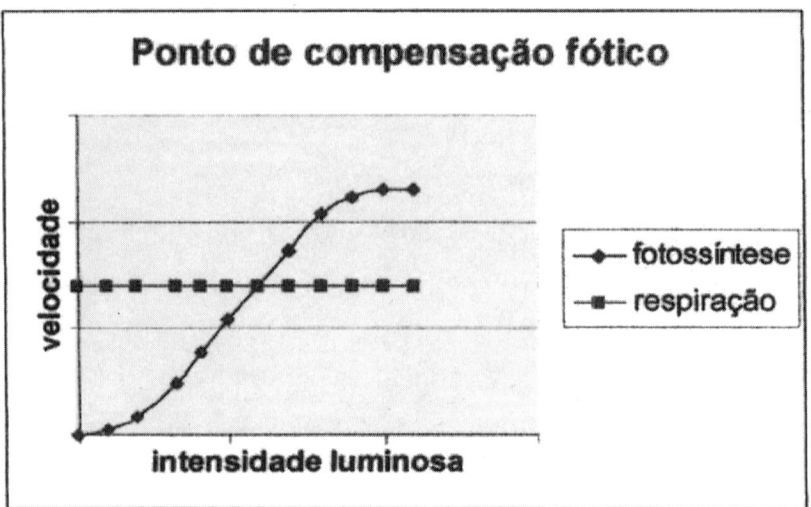

Figure 6-2. A graph representing velocities of respiration and photosynthesis. "Ponto de compensação fótico", "fotossíntese", "respiração", "velocidade" and "intensidade luminosa" mean: compensation point, photosynthesis, respiration, velocity and light intensity, respectively.

At one point during the discussion following the oral presentation of the chloroplasts group, the teacher drew a graph on the blackboard similar to the one shown in Figure 6-3. The graph resembled the students', except that it substituted time for sunlight intensity.

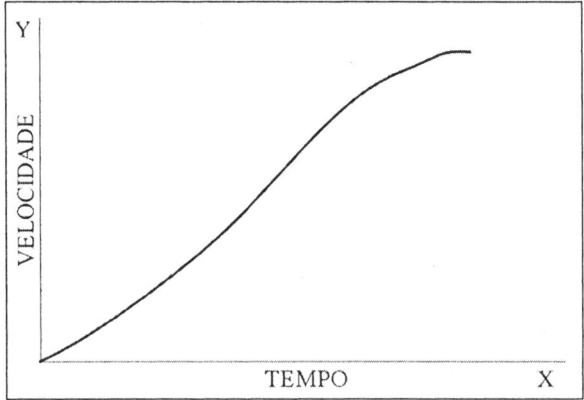

Figure 6-3. The teacher's graph on the blackboard.

The teacher argued that sunlight changes over time, and so one could think of the graph drawn by the teacher and the students' graph of photosynthesis as having the same shape, if one considers an interval of time in which the sunlight is increasing with time. Then the teacher asked them how they would evaluate the "total amount of photosynthesis" at the end of, for example, five hours. Many different ideas came up, but the discussion did not evolve. The teacher tried a different strategy, asking them what would be the total distance traveled by a car if the graph were of velocity of a car versus time. Quickly a student proposed that the area under the graph would have to be evaluated, probably drawing on previous experience in high school with this type of physics exercise. The result was applied immediately to the photosynthesis graph, and the teacher asked them how they could evaluate the area under the graph, since the graph was not a straight line. No response was offered, and the teacher drew a graph like the one in Figure 6-4 saying that the sum of the areas of each rectangle would be an approximation of the area under the curve.

This episode ended as the teacher asked how they could approximate the area more closely, and a student answered that one could make the bases of the rectangles smaller and smaller. This solution was probably inspired by their previous experience in this course with a task developed through the experimental-with-technology approach that aims to introduce the notion of derivative. The task was posed as follows: Is it possible to make the graph of

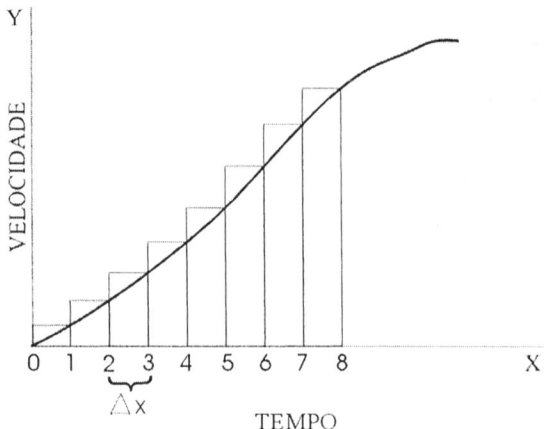

Figure 6-4. The teacher's proposal to calculate the area under the graph.

parabola with secant lines joining points $(x_1, f(x_1))$ and $(x_2, f(x_2))$ on the curve and then made the intervals $[x_1, x_2]$ smaller and smaller to obtain better approximations of the parabola. They finally concluded, with the aid of the teacher, that lines tangent to the curve at each point should be sketched to obtain the parabola, since tangent lines are the best linear approximation of a curve at each point (Borba and Villarreal, 1998). But how can the 'modeling experience' of this group be interpreted from the theoretical perspective that we are presenting here?

Of course we could focus on the role of the teacher in the process where technology and/or modeling is involved, as has been the case in many different studies throughout the world, as well as in our research group and graduate program (e.g. Barbosa, 2001; Penteado, 2001); or we could focus on the mathematics produced by the students, as one member of our research team has done (Malheiros, 2004). Without overlooking either of these, our focus will be on how the humans-with-media collective, formed of students, teachers and different technologies of intelligence, acts within this pedagogical approach. In the episode presented above, human as well as non-human actors were described that shaped the project, its results, and the teaching that was developed by the teacher. There were blackboard and chalk (and the student version of these items, paper and pencil), graphing calculators (although they were not used during the presentation, and there is no evidence of direct use in their work), the *Excel* software and the books used in their research. The software seems to have been used in two ways. At times it served simply as a tool for drawing; a collective was formed, and the way they found to reproduce a given graph that they found in the book was using *Excel*. But once *Excel*, or other similar spreadsheets, come into play, new paths become possible. For example, the students used the algebraic facilities of the software to experiment and come up with the equation that modeled the multiplication of chloroplasts vs. time. They could have done this without using *Excel*, but what matters in our discussion is that they were able to use a software to plot the information they found in books (the chloroplast divides itself into two every 20 minutes) and to test the equation they found $(f(t)=2^{(t/20)})$. The teacher helped, especially with the equation, and there is a clear interaction with other activities developed in the class, as they were accustomed, for example, to studying the relation of different coefficients of families of functions (e.g. linear and quadratics) with their graphs. They also used the book to study the logistic model, and showed how much and how little they were able to learn independently, which is not common among students just out of high school in Brazil. There is no claim, therefore, that only the software shaped the mathematics developed by the students. In line with the general theoretical discussion developed in Chapter 2, a new medium does not erase or suppress the previous one, especially since the others were so much a part of the students' educational experience. The knowledge they produced is an example of the kind of mathematics that is constructed by collectives composed of humans, math and biology textbooks, *Excel*, and pencil and paper in the context of a math course for biology majors.

4. MODELING WHEN THE INTERNET BECOMES AN ACTOR

Information technology (IT) artifacts like *Excel* or *Cabri II* can already be considered 'old' new IT; old, in the sense that they have been used by many students, and still new in the sense that professors have yet to incorporate them much into their teaching activities. The reason for this is that, since this first wave of technology has been incorporated into collectives that produce knowledge, another wave that includes the Internet has transformed the notion of work in a way that we have not yet fully understood. It seems that we are 'surfing' on different waves of technology of intelligence which are changing dramatically the way humans-with-media know. More and more people have access to the Internet, whereas few students and teachers know how to deal with software of the first wave. Similarly, some students are not very familiar with libraries, and feel more comfortable with Internet searches than with library research. Information and communication technology (ICT), libraries, calculus software, and *Excel* are all shaping the projects developed by the students in the classes for biology majors we have researched.

Another group of students, in 1997, almost gave up their investigation of the theme 'abortion' as they could not find information supporting both sides of the issue ('right to choose' vs. 'right to life') that went beyond mere 'propaganda'. Eventually, with the help of the teacher, they were able to develop their project. It is fair to suppose that, if the project had been developed today, they could have found many home-pages on the subject. Indeed, in an analysis of the 15 projects developed by the two classes in 2001, more than two thirds have used only, or mainly, the Internet as their basis to start their project. There is strong evidence of this, such as students' comments in their oral or written presentation, as well as weaker evidence, like the style of the text presented, and figures that appear to have been downloaded. Out of the other third of the projects, there are two about which we can make no claims regarding use of the Internet, two for which the Internet does not appear to have been used, and one in which the Internet was used but was not the main source. In any case, there is an apparent trend for projects to have *webographies* instead of bibliographies - and this was the case, even for the 'more mathematical' project that was developed regarding fractals. Since there were no books on the subject in the library which were accessible to the students at the time, the teacher lent them a book (Devaney, 1990) and suggested a specific chapter to study. They did study the book, but soon their main resources became the teacher, who explained the notion of fractal dimension to them; a teacher from the biochemistry department, who showed them a nice example of application of fractals in his field; and the Internet, where there are examples of fractals in many sites.

A more typical project, also from 2001, was about 'mad cow' disease (*Bovine Spongiform Encephalopathy* - BSE). There are probably few books published on this theme, as it is a relatively recent issue and, until the early 90's, it was not important. Since Canada and Brazil were engaged in a dispute involving the sale of airplanes, and the former decided to suspend the purchase of meat from the latter, it became a very hot issue in Brazil because it was understood as a retaliation, despite the fact that there has never been a single case of this disease in Brazil due to the type of cattle feed. In spite of the dispute between the two countries, which associated airplanes with mad cows in a amusing fashion, generating thousands of jokes, the students were interested only in understanding the mad cow disease from a biological point of view.

In the final written version of their project, they presented a history of similar diseases and how the mad cow disease differs from them. Their 'bibliographical resources' were science magazines, Newsweek, and web sites in which they also found the statistics for new cases per year from 1987 through 2001. See graph in Figure 6-5:

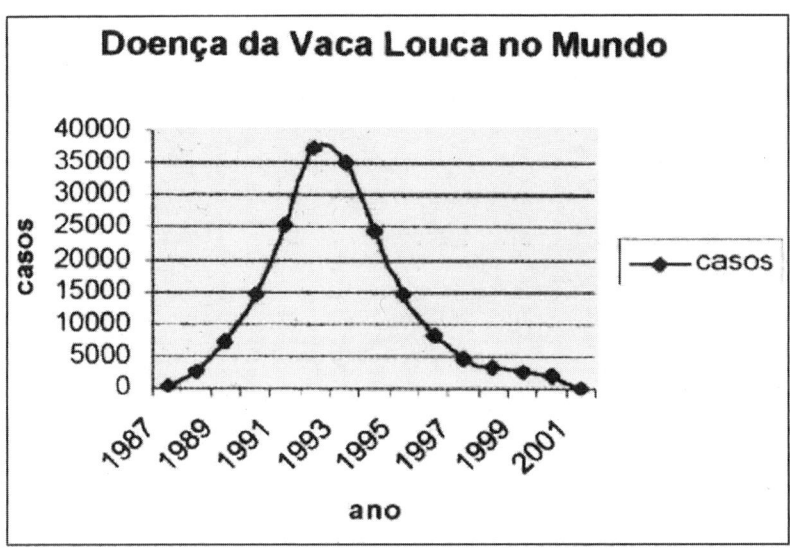

Figure 6-5 . Cases of mad cow disease in the world. "Casos" and "ano" mean cases and year, respectively.

Although the teacher noticed that the data for the year 2001 was not very accurate, as the year was far from finished, there was a clear pattern, and the teacher posed a challenge for them: to describe algebraically what was happening and see if, using either algebra or the graph, they could predict what was likely to happen. As shown in Figure 6-5, the students were using *Excel* to plot data and to solve the problem posed to them. They used different *Excel* commands to come up with a polynomial function of the 6th degree. The teacher, during their oral presentation, told them that the function was a good descriptor of their data, but perhaps not very appropriate for predicting the trend. There was quite an intense debate in the classroom, but another issue arose, which had been raised previously by the teacher with the students during office hours. The students had followed the recommendation of the teacher and had generated a graph in which the y-axis presented the number of cases cumulatively instead of by year (see Figure 6-6).

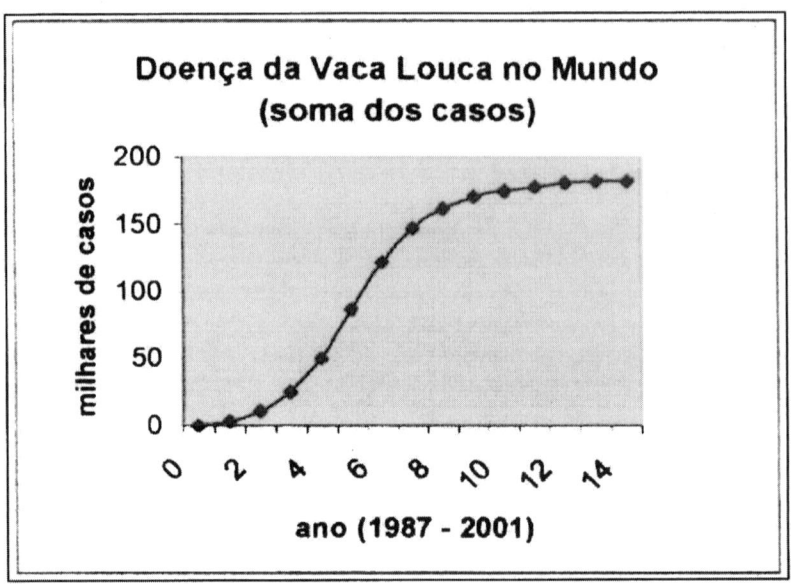

Figure 6-6. Cumulative number of cases of mad cow disease in the world. "Milhares de casos" and "ano" mean thousands of cases and year, respectively.

They also studied, independently, the section in the textbook about the logistic model and presented the graph in Figure 6-7 to the classroom.

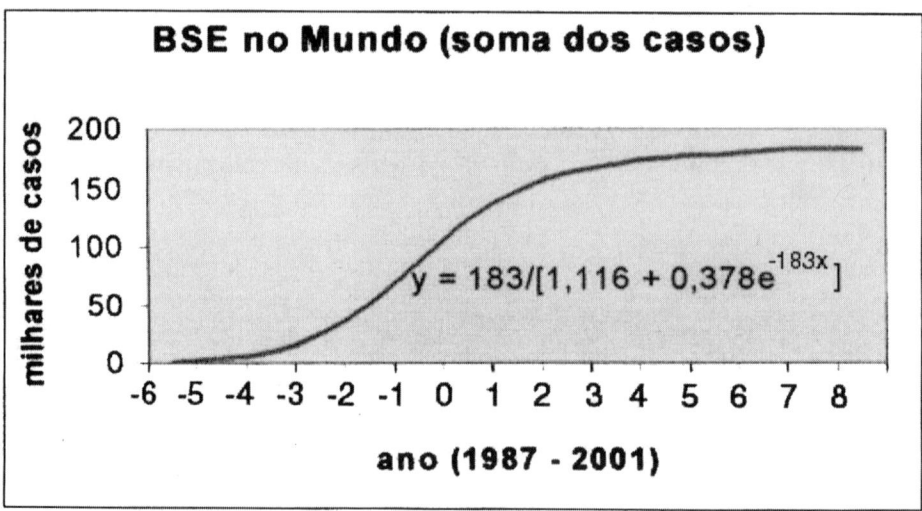

Figure 6-7. A logistic model for mad cow disease. "BSE no Mundo (soma dos casos)" means BSE in the world (total number of cases).

After the students explained that the *−6* on the x-axis was 1987, and that they had done it that way so it would fit in the algebraic model they found in the book, the teacher challenged them to generate a graph and coordinate it with an equation, in which the zero on the x-axis represented 1987. After a period of discussion, the group built on an idea that emerged from the other students in the class and, drawing on activities they did during the experimental-with-technology part of the course in which they researched the coefficients of quadratic functions, they came up with the graph and equation shown in Figure 6-8.

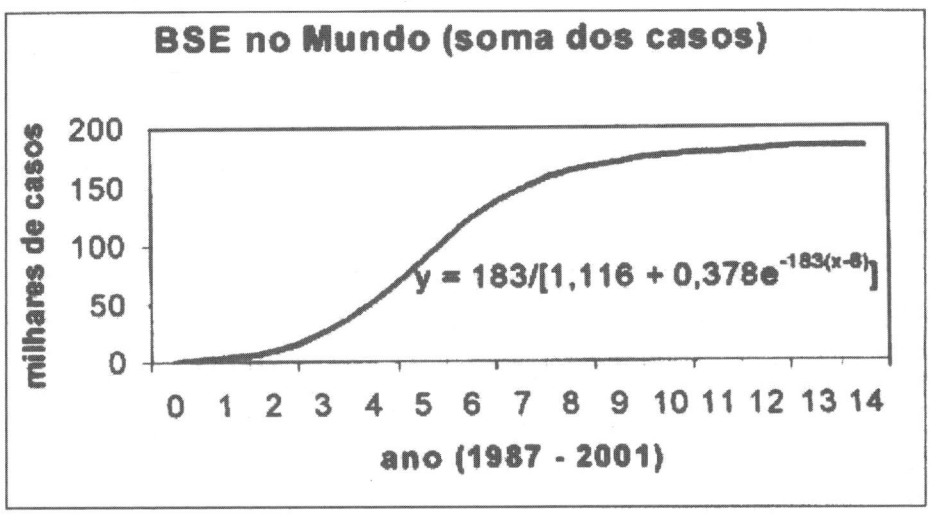

Figure 6-8. A new logistic model for mad cow disease.

As was shown in further analyses conducted by members of GPIMEM, there were discrepancies between the equation and the graph they presented that were not noticed by the teacher at the time. In Malheiros (2004), there is a more complete analysis of the possible mistakes in the equation that, once corrected, would resolve most of these discrepancies. However, the role of the teacher, and possibilities of what could have happened, are not the focus at this point.

It is important to notice that their experience with the graphing calculator was still present in the classroom at the time, despite the fact that there were none being used. In our perspective, the graphing calculator was also an actor in the graph situation above, as a result of their interaction with this medium in other problems, just a few classes prior. We would claim that the ideas developed in the above example were generated by a humans-with-Internet-graphing calculator-*Excel* collective. The role of the graphing calculator and *Excel* in their production is quite clear, as they were active in generating graphs and equations. The role of the Internet did not seem to have had 'mathematical' aspects in this interdisciplinary approach to curriculum. It is quite clear that, as their main source of data, it enabled them to develop the project but we still do not have a good example in which the Internet has 'shaped' the mathematics produced in the classroom more actively; or it could be that our researchers' eyes are not prepared to see it, just as some thought that computers would not change mathematics teaching and learning at all, back in the beginning of the first wave of computers in education.

Although our research group has reached no conclusions regarding how the Internet has shaped face-to-face education, and some of us are deep into exploratory study in this theme, it can be noticed that, in the case described above, the Internet seemed to bring more interdisciplinary aspects to the investigative process associated with mathematical production, which up until recently, has been unusual in most undergraduate education.

5.   MODELING , HUMANS-WITH-PAPER-AND-PENCIL AND... POTATOES

The reader may have noticed that there have been no demonstrations in the examples shown, even if one were to be more flexible regarding proof and not require it to be strictly formal. We also found no trace of it in the more than one hundred projects analyzed by the GPIMEM team involved in this part of our research. This could be due to the fact that the students are biology majors; it could be because the teacher does not stress this aspect of mathematics in his teaching (although he does demonstrate results to students), it could be due to the maturity of the students, or to the fact that the modeling approach has other criteria for satisfying one's certainty about whether something is correct. Although, we personally believe that all of these offer clues for a more global answer, we would like to concentrate on the last one.

Pierre Lévy has claimed that the introduction of computers could change the very idea of truth, seen as resulting from theory, to a more pragmatic view of truth, based on simulation, in which models are generated which are considered to be provisory, and are evaluated using criteria of relevance, efficiency and usefulness. One could compare these more pragmatic criteria of truth to mathematical modeling and experimental mathematics, wherein if a model works to describe and/or predict, it is accepted, even though it may not always be firmly based in theory. Fractals and fuzzy differential equations could be examples of mathematics that have been used in different areas without a formal foundation from a strict mathematical point of view, like analysis was to calculus. The combination of the modeling approach with the use of IT or ICT tends to bring these criteria of truth to the production of mathematical knowledge in the classroom. We by no means want to generalize and say that this is the kind of mathematics that emerges from a class in which modeling and computers are welcome. As mentioned before, there are various other factors that shape this educational environment; and it is certainly possible that no demonstration would have taken place (on the part of the students) if exercises had been proposed within a more traditional approach in this class. Demonstration was, in fact, taking place, in a broader sense, during the part of the class where students conjectured about the relationship between coefficients of functions and graphs.

For the purpose of this book, we just want to raise the idea that new paths have been opened up for arriving at socially acceptable results in the mathematics classroom, and this is related to the new ICT actors that joined collectives that produce knowledge in the classroom. In the so-called service courses, in which mathematics is in service to other subjects like biology, there may even be a trend to omit demonstration completely.

It is important to point out that the proof, even in the mathematics community, is associated with the audience (Thurston, 1994). In our case, the audience is biology majors, for whom rigorous mathematical proofs seem not to be necessary, and sometimes "visual proofs", word argumentation or classroom debates are enough to justify some assertions. The students believe in the teacher's statements or in the books, which are considered to be authorities. Even in courses where some proofs are done and the teacher asserts that "everything must be proved in mathematics", there paradoxically exists a kind of belief, a set of truths whose validity is not questioned, and the course goes on without proving them. Why should mathematical proofs be important for a biology student? They just use the mathematical results; they don't worry about the proofs because they know someone else has proved them. We must remark that researchers from other disciplines do the same.

We believe that modeling projects could generate mathematical statements that merit proofs, however the focus in this approach is not on the proof. There are other points in the class that would be more appropriate for developing that activity, such as when mathematical conjectures arise in the experimental-with-technology approach. As will be seen in the next example, this does not mean that if computers are not used, demonstration and algebraic reasoning will predominate.

In one of the first modeling projects in the biology course, in 1993, neither graphing calculators nor computers were available for the students. The students were concerned with overpopulation and how the excess of humans could be a problem for the planet and for humanity itself. Like their counterparts in later courses, they generated graphs and established projections for data that could be modeled with functions, such as "approximately one million people are born every five days", and "the earth, which is home to five billion people today, will hold another billion by the year 2000". Unlike the other groups, no algebraic expressions were proposed to model their population data and predict future population. Again, we cannot jump to conclusions, as there are many factors involved, but it is reasonable to say that, if graphing calculators or computers with plotters had been available, the chances of them coming up with an algebraic representation of their data, or a Cartesian graph instead of a bar graph, would have been greater. So it could be the case that the mathematics produced by collectives of humans-with-paper-and-pencil has this characteristic in a setting like the biology classes, but it could have other characteristics, as described in the following example.

One of the most famous examples of Brazilian mathematics education comes from a classroom in which the modeling approach was being used. Although we mentioned it in Chapter 3, we would like to expand on it at this time. The example has no explicit connections with computers, which were uncommon in the mid 80's when the scene took place in one of the classes taught by Rodney Bassanezi[19] (see Bassanezzi, 1994), one of the first teachers to use this teaching approach in mathematics classrooms in Brazil. He was assigned by the mathematics department of one of the major research universities in the country, UNICAMP, to teach Calculus I to food science majors. As he entered the classroom, he noticed that many students were wearing T-shirts printed with the phrase *I hate calculus*. Students told him that they did not consider calculus to be useful for them- an opinion probably shared by many students in different parts of the world. The teacher proposed to the students that he would work with mathematics using their own problems as a starting point. Many themes arose, but a question posed by one of the students set the stage for a project that involved the whole class: "My father plants potatoes, placing each seed 30 cm apart; I would like to know why he does it this way".

The first thing they did was to find some data in the agriculture department. They found out the average number of potatoes per plant (eight), the distance between each row so that weeding is possible (80 *cm*), the average weight of eight potatoes (639 grams), and that one 'alqueire' (24200 m^2) should yield eight hundred 60 *kg* sacks of potatoes. They also found out, based on some empirical agricultural research, the average number of potatoes as a function of distance between plants:

---

[19] For more details on this example see Gazetta (1989)

25 cm → 4,5 potatoes
30 cm → 6,5 potatoes
35 cm → 7,5 potatoes
40 cm → 8,0 potatoes

After they collected all this data, the student's question was transformed into "Find the distance between plants (in the same row) so that production is maximized".

They followed up on their modeling activities by first considering a square area of one 'alqueire' to plant potatoes, and concluding that that area would have 194 rows of 155 *m*. each. Then they generated a function that represented the production (*P*) of one alqueire in number of sacks as a function of the distance between plants (*d*) and the amount of potatoes per plant (*p*).

P(*d,p*)= (average weight of one potato × number of potatoes per plant) × (number of plants per row × number of rows)/(weight of one sack)

$$P(d,p) = \left(\frac{0,639}{8} \cdot p \right) \cdot \left(\frac{155}{d} \cdot 194 \right) \cdot \left(\frac{1}{60} \right) = \frac{40 \cdot p}{d}$$

The next step was to express *P* as a function of just one variable. As they had data for the number and average weight of potatoes as a function of the distance, they first created a continuous function

$$p = f(d) = 8,5 - 2^{(7-20d)}$$

and finally arrived at

$$P(d) = \frac{40}{d} \cdot (8,5 - 2^{(7-20d)})$$

Afterwards, they found the derivative of the function using the derivative rules, obtaining

$$P'(d) = \frac{40 \cdot 2^{(7-20d)} \cdot (20 \cdot \ln 2 \cdot d + 1) - 40 \cdot 8,5}{d^2}$$

and then, made $P'(d)$ equal to zero to find the optimal point. As the solution for this equation was not simple, they constructed a table for $P'(d)$ and used a numerical method, called bisection, and the Mean Value Theorem to solve it. They concluded that the optimum distance would be 31,5 cm, which would yield 873 sacks, and after they included the minimum number of sacks of potatoes per alqueire one has to produce in order to cover costs (800 sacks), they solved the inequality $P(d) \geq 800$ to find an interval that ranged from 27 cm to 40 cm. This inequality was solved numerically and graphically. In the first case, the bisection method was used after first doing some algebraic calculations to obtain a simpler expression

$$800 \leq \frac{40}{d} \cdot (8,5 - 2^{(7-20d)})$$

$$2^{(7-20d)} \leq 8,5 - 20d$$

$$8,5 - 2^{(7-20d)} \geq 20d$$

or

$$p \geq 20d$$

The graphical solution of the inequality required finding the values x_1 and x_2, shown in Figure 6-9, but first it was necessary to study the domain of validity, asymptotes and roots of function P, apart from the maximum that was determined earlier.

This example illustrates how collectives of humans-with-paper-and-pencil produce mathematics in a classroom where modeling is the pedagogical approach being used. We have no research data for this example, only reports of it, so it is not possible to analyze how they developed their mathematics in the classroom, nor the role of the teacher. It is possible to say, however, that when the students were tested at the end of the course, only one out of seventy students failed, which was a far better result than in previous years. Gazetta (1989) describes in her masters thesis how, in the process of developing the potato project throughout the course, the students took many 'detours' to study exponential functions, inverse

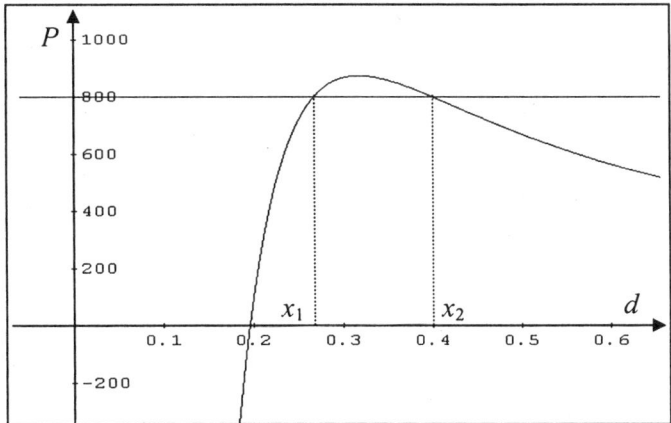

Figure 6-9. The graph of potato production (P, number of sacks) per alqueire vs distance between plants (*d*).

functions, derivative of these functions, continuity, asymptotes, roots of functions, the Mean Value Theorem, and other topics. They studied these topics as the need arose to solve the problem that involved, at least, most of the class. The curriculum thus followed the students' and the teacher's interests, conciliated with the teacher's 'ulterior motive' of following the syllabus so that the students would be able to pass the exam. So, although we cannot claim that modeling was an effective way to learn because the students did well in the test (but only that the students also learned how to solve test questions), we believe that experiences like the ones reported here tend to support the idea that there are options for curricula which are not based on the way scientists organize their knowledge, in our case the way mathematics is often seen as a building that is the result of a sequential construction process. The two examples presented in this section lead us to consider aspects related to the structure of mathematics curriculum when modeling is used as a pedagogical approach and there are no ICT actors involved in the construction of mathematical knowledge.

6. SCIENCE IN ACTION IN THE CLASSROOM AND VIDEO CLIP CULTURE

Modeling proposes an organization of curricula that distinguishes itself from the rigid and sequential structure that characterizes the traditional school mathematics curriculum. Many believed that mathematical content, organized into a deductive system that emphasizes logical order and the axiomatic method, should be translated into mathematical curriculum[20]. This was the case of the new math movement in the 60's, or, as proposed in the critique by Moreira and David (2003), the case of the didactical transposition, in which the structure of scientific knowledge is left untouched. Many theories that dichotomized scientific concepts (high level) and everyday concepts (informal and low level) were developed, and are still very strong within the mathematical education community.

The modeling approach also challenges the *exercise paradigm*[21], bringing some research practices into the classroom as students work with different kinds of uncertainty regarding their data: students do not know where to start, and once they are able to identify a topic to investigate, it is usually not organized the way data are in textbook problems. Students are forced to define their own problem, which is one of the main challenges in many scientific fields today, which are already crowded with papers and books. Modeling, inherently, also breaks the rigid barriers between disciplines, as different groups of students are not likely to choose problems from the same field. Interdisciplinarity becomes common, and transdisciplinarity may be seen as a goal to aim for.

For many of us who are frustrated with the way students see mathematics, even those who do not use T-shirts declaring their distaste for the subject, opening the door for students to participate in the curriculum becomes a possible path to overcoming such aversion. Opening up curricula to allow for different kinds of 'mathematical participation' in the classroom is a key feature of the approach presented in this book. But all is not rosy within modeling, as, for example, when we invite students to choose a theme of their interest and they fail to do so. There are several cases within the more than one hundred cases analyzed by our research team where the evidence is quite strong that they did not choose a theme truly of interest to them, but rather an easy one, or a topic they felt would please the teacher. In Borba, Meneghetti and Hermini (1999), we reported a case we believe to be typical. There is strong evidence that a group of students was not engaged in the theme they chose. The topic 'Devastation of forests in Brazil' was popular at the time, especially for biology majors. However, although it was suggested, very little math was used, and there was no sign that the students actually became engaged in the topic. Araújo (2002), a researcher associated with GPIMEM, who developed her research in a math course for engineering majors, also found that students may choose problems, even when given the choice, for reasons that differ significantly from the ones

---

[20] For a discussion about this issue see Kline (1970), Kaput (1994) and Alrø and Skovsmose (2002).

[21] For a detailed description of the *exercise paradigm* see Alrø and Skovsmose (2002).

researchers like ourselves would like to believe. Students may be too influenced by their previous educational experiences, and may thus tend to choose problems that resemble textbook problems.

This is not the only problem. As Barbosa (2001) has analyzed indepth, it is a challenge to be the teacher implementing the modeling approach; and as he also points out, it is not an easy task to prepare future or pre-service teachers to apply the modeling approach in teacher training programs where modeling is not taken seriously. Continuing education for teachers is not always successful, as Anastácio (1990) has documented, although this is the educational setting where modeling has been most popular. There is no research to our knowledge, however, documenting what teachers who have participated in such courses have done afterwards in the classroom.

Another open problem emerges when students choose a mathematical topic as their theme of investigation. In the example mentioned briefly earlier in this chapter, a group of students chose fractals. While the notion of fractal dimension is accessible to students at the level of first year biology majors, it is not so simple that the teacher felt comfortable simply referring students to some book, as he did in the case of the logistic model. In the fractal example, the teacher gave them a reference (in English at the time, which greatly augmented the difficulty of first year students). Some difficulties arose, and the teacher (and perhaps friends from other disciplines whom they may have consulted) helped them. Having them read the book was one teaching strategy. The teacher prepared a didactical sequence, first relating integer dimension to number of copies as a means of introducing the notion of non-integer dimension, or fractal dimension, as proposed by Devaney (1990). The notion of non-integer dimensions is fundamental for fractals, making it necessary, in this case, to devote some time to discussing it using a more traditional didactic approach; thus, one cannot say that learning and teaching in modeling involves using only an open, unstructured approach.

So, if a group chooses number theory, or some topic associated with it, thus demonstrating, perhaps, an inclination to be mathematics majors, there are still pedagogical issues to be solved. In other words, although GPIMEM conducts research on modeling, and values modeling as a pedagogical option, we recognize that, even if it were implanted on a large scale, as opposed to just one mathematics course in a biology program, it is very likely that it would transform and incorporate other pedagogical approaches rather than doing away with them. Problem solving, ethnomathematics, reading, traditional teaching would probably not be suppressed but would instead gain new colors, as the whole chain of curricula would not be some kind structure based on a fixed order of science. Lévy (1993) proposes that a new medium does not eliminate the other, e.g. computers will not supplant paper and pencil. Similarly, we can propose that modeling does not eliminate other approaches. In both cases, then, a new medium or a new pedagogical approach transforms the other employed together with it. A feature that searching on the Internet, modeling and the experimental-with-technology approaches have in common is that, in all of them, students have to work with abduction, (Shank and Cunningham, 1996) as we briefly discussed in Chapter 4.

Previously we presented a matrix (see Table 3-1 in Chapter 3) illustrating perspectives regarding the role of technology within different approaches to modeling in the classroom. The role of information and communication technology could be understood in terms of a spectrum ranging from tutoring, simply providing answers, to becoming an important actor in investigations conducted by collectives of humans-with-media. Such interaction leads to reorganization of thinking as we have proposed, where technology is simultaneously inside and outside of us. Access to ICT is seen as a right and as a possible means for democratic education. Conceptions of modeling, in turn, also vary over a range that includes: mere applications, as in textbooks; new topics for new courses; projects in which the theme is chosen by educators; or in the way highlighted here, which stresses greater participation of students, not only in taking different problems down new paths, but also in choosing problems. The exploration of problems using the modeling approach, and experimentation with software that have graphical images or html colors, connects one to other 'worlds' beyond academia.

Such ideas and activities are more in line, and have synergy, with the video-clip culture that is growing up alongside the generations raised with TV's that have remote controls, making 'channel-surfing' possible, and where video clips with impressions, with fast-changing images, with no clear relationship between image and discourse, are becoming the norm. Modeling has synergy with this 'culture', as different students do different things and can learn pieces of mathematics as needed for something they are interested in, until (if so) mathematics becomes their main interest. 'Traditional software' of the first wave got closer to this with their multiple options, multiple windows and so on, but also because they became co-actors of different investigations, for different students who prefer graphs, tables, algebra or other representations (Borba, 1993).

Second wave ICT, marked by the Internet, is even more similar to this 'culture', and it may explain part of its attraction. WWW is colorful and more and more similar to video-clips, as new interfaces are developed that allow for pop-up windows, sounds, color, smell, different options, and possibilities of navigating from a boring page to a more interesting one with just one click. Browsers have more in common with remote controls than one may think. And if Kerckhove (1997) is correct, these new kinds of texts (labeled hypertexts and hypermedia), which are in the process of being created, strike us at an emotional level before they reach the level of conscious reflection, and structure our brain. This is a possible explanation for why so many people are attracted to the Internet by its colors and its hypermedia text: because it is part of the TV culture that is a part of each of us, at least to some extent, and because we can use it at work, at home, in school. We hope that modeling and this synergy with the new interfaces transform universities and schools into places where investigation and new texts transform the way we see books and the very nature of teaching and learning.

Chapter 7

EXPERIMENTATION, VISUALIZATION AND MEDIA IN ACTION

1. INTRODUCTION

Using some examples of modeling projects at the university level, we have shown, in Chapter 6, how collectives of humans-with-media learn mathematics when modeling is used as a pedagogical approach, and how this alters mathematics curricula and the role of the teacher. In this chapter, we would like to present some examples, stressing visual or experimental aspects in each of them, and discussing the relation of those aspects with media and the reorganization of thinking.

The examples in this chapter come from projects involving classroom research, interviews with students, analysis of excerpts from mathematical textbooks, and teaching experiments. Although different research methodologies used by researchers of GPIMEM and associates will be discussed later, it should be noted that *teaching experiments* in this book refers to a sequence of meetings with one or two students in which a teacher-researcher builds models of students' thinking regarding problems posed by the students or the teacher. This type of procedure is based on the work of Steffe and Thompson (2000), and Cobb and Steffe (1983), and our version can be studied in detail in Villarreal (1999) and Benedetti (2003).

2.   EXPERIMENTING WITH PARABOLAS: VISUAL CONJECTURES

The following episode comes from the 1998 applied mathematics course for first year biology majors at the State University of São Paulo (Rio Claro - Brazil)[22]. The students were working with graphing calculators, and the task assigned was to investigate what happens to the graphs of the quadratic functions $y = ax^2 + bx + c$ when the parameters a, b and c vary. Students were free to take different paths in their investigations, which they did. After the students worked on this task, the professor led a discussion to systematize the students' findings. One of the groups came up with an original conjecture that was stated by Renata in the following way:

> When b is greater than zero, the increasing part of the parabola will cross the y-axis . . . When b is less than zero, the decreasing part of the parabola will cross the y-axis. [The student gestured in the air with her hand to illustrate]

This statement caused an intense discussion in the classroom. Neither the teacher nor the other students had thought about b before in this sense, and no one was sure whether the conjecture was true or not. After a lengthy analysis, they concluded that the conjecture was true. A summary of the path the discussion took can be understood if we analyze each of the four possible combinations of the signs of a and b in the vertex formula $x_v = -b/(2a)$. We can see that, considering the position of the parabola's vertex on the coordinate plane, and the concavity of the curve, we can confirm the conjecture. For example, if $a>0$ and $b>0$, then $x_v<0$, and since the parabola is concave upward because $a>0$, we can have any of the graphs in Figure 7-1.A, which show that all parabolas intersect the y-axis with its "*increasing part*". Analogous reasoning may be made with the other cases to see the validity of the conjecture made by Renata and her group.

We used the word 'see' when we talked about the validity of the conjecture because it can be demonstrated through a 'visual proof'. Some weeks later, another way to prove the conjecture arose, when the students were studying the concept of derivative associated with the intervals where a function is increasing or decreasing, points of maximum or minimum, etc. This time the teacher presented the following proof: the derivative of $y = ax^2 + bx + c$ is $y' = 2ax + b$, and when $x = 0$, $y' = b$, which means the function is increasing at $x = 0$ if $b>0$, or in Renata's words, "*the increasing part of the parabola will cross the y-axis when b is greater than zero*". For $b<0$, the teacher did an analogous proof.

---

[22] The course was taught by Marcelo Borba, as described in Chapter 6.

From this example we can see that the calculator was used to plot numerous graphs, which inspired the students to formulate an initial visual conjecture. In the debate with the whole class, there were some attempts to make visual arguments to support the conjecture. Some weeks later, the teacher provided a new justification (using mathematics he had just taught) that reinforced the validity of the initial conjecture.

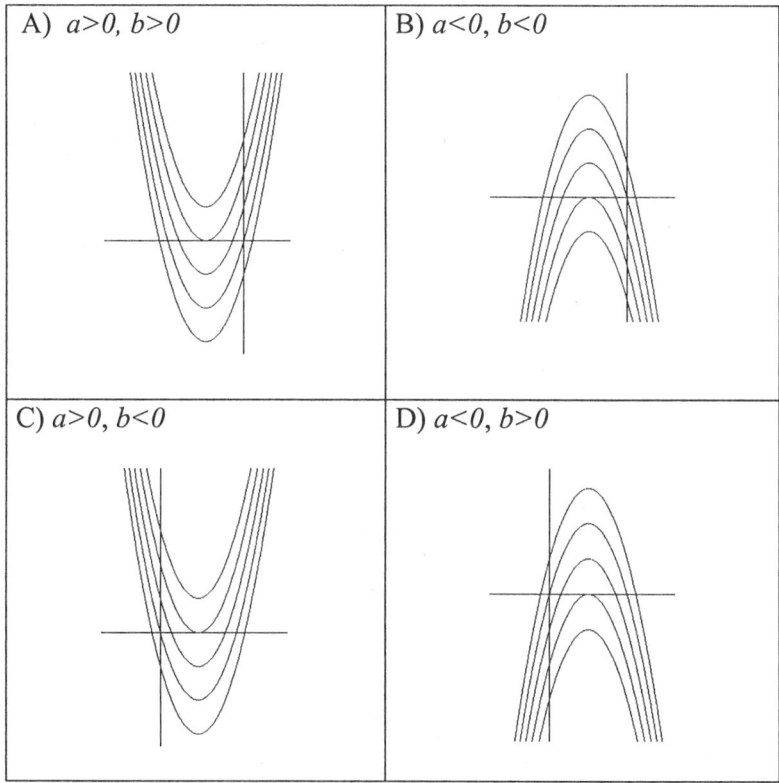

Figure 7-1. The four possible combinations of the signs of *a* and *b*.

The parabola activity provoked the emergence of an original conjecture that provides, for example, a new way to decide the sign of *b* from looking at the graph of the parabola, or more information to sketch the graph of a parabola knowing the sign of *b*. We may think about the value of Renata's conjecture from a mathematical point of view, but this is not a central point; what is really important is the engagement of the students in the process of conjecturing and discovering a mathematical result unknown to them. Also, it demonstrates how teachers can make room for group conjectures and sharing some of the conclusions with the whole class.

The same open-ended task was presented to the 2002 mathematics class for first year biology majors. Different groups tried different equations with the graphing calculator and generated written reports for the class discussion. One student plotted the graphs of equations $y = ax^2 + bx + c$, maintaining the values for b and c constant and changing only the value of a. Then she came up with a conjecture: "*when a increases, the roots decrease*". The graphs in Figure 7-2 show what she meant: the roots of the parabola with the greatest value of parameter a would be between the roots of the parabola with the smallest value of a.

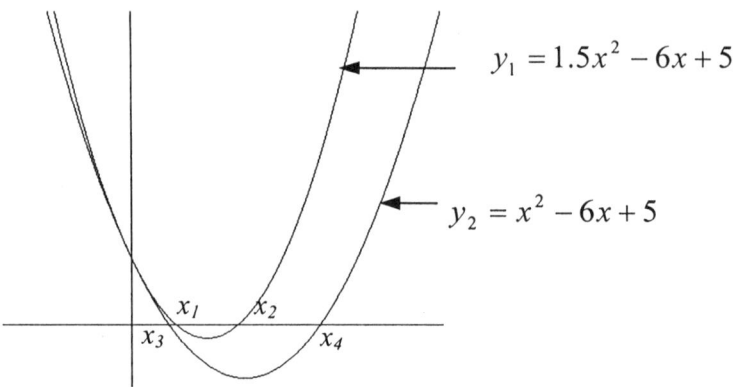

$$y_1 = 1.5x^2 - 6x + 5$$

$$y_2 = x^2 - 6x + 5$$

Figure 7-2. Two parabolas showing the student's conjecture.

This conjecture was not discussed very much in the classroom; some speculations were raised about whether the conjecture was true or not, but there was no proof or informal evidence offered regarding its validity. The teacher tried to raise the issue again the following week, but was unable to engage the class. However, the topic attracted the attention of the teacher and an undergraduate student in mathematics, Ricardo Scucuglia, who was developing a research project. Ricardo watched the videotape of the discussion in the classroom and decided to study the truth of that conjecture. In a written paper presented at a research conference for undergraduate students, he stated:

> It was noted that the student's conjecture was coherent, although she did not express it correctly. Based on this analysis, I state her conjecture: Let $y_1 = a_1x^2 + bx + c$ and, $y_2 = a_2x^2 + bx + c$, a, b and $c \in \Re$ be equations that represent second degree functions of $\Re \rightarrow \Re$ with $x_1 < x_2$) roots of y_1 and $x_3 < x_4$ roots of y_2. Prove that if $a_1 > a_2$, then $x_3 < x_1$ and $x_2 < x_4$. (Scucuglia, 2002)

Ricardo wrote the conjecture in mathematical language and decided to prove it. In the process, he realized that he needed additional hypotheses for the conjecture: 1) $c \neq 0$, since if $c = 0$, the two quadratic functions would have zero as a root, and the thesis would not be true, and 2) $a_1 > 0$ and $a_2 > 0$, since if they were negative, the relation between the roots would be inverse, and if a_1 and a_2 had opposite signs, the conclusion would not be true. With those new hypotheses, Ricardo reformulated the conjecture:

Let the functions $\varphi_1 : \Re \to \Re$, $\varphi_1 = a_1 x^2 + bx + c$ and $\varphi_2 : \Re \to \Re$, $\varphi_2 = a_2 x^2 + bx + c$, a_1, a_2, c be real nonzero constants, b real constant, x_1 and x_2 roots of φ_1, x_3 and x_4 roots of φ_2, with $x_1 < x_2$ and $x_3 < x_4$. Prove that if $a_1 > 0$, $a_2 > 0$ and $a_1 > a_2$, then $x_3 < x_1$ and $x_2 < x_4$.

Finally, Ricardo did prove the conjecture considering different cases: $c > 0$ and $b^2 - 4ac > 0$ (this last condition ensures the existence of roots); $c < 0$ and $b > 0$; $c < 0$ and $b < 0$; $c < 0$ and $b = 0$.

In this case, a conjecture that arose during an exploratory activity with biology students, in a collective of humans-with-graphing calculator, created the environment for an investigation by a mathematics undergraduate student, who engaged in the task of reformulating and proving the conjecture. This environment could be considered an extension of what Skovsmose (2000) calls the *landscape of investigation* with reference to pure mathematics[23]. Ricardo was investigating the truth of a mathematical conjecture raised by a biology student in a particular mathematics course, and that situation generated a peculiar landscape of investigation for him, even though he was not enrolled in the mathematics course.

Although the formulation and validation of a conjecture were not relevant activities for biology students who were engaged in mathematical experimentation with graphing calculators, they nonetheless provoked the interest of a mathematics student. This particular example also shows different roles of actors in an 'extended collective' of humans-with-media that included a junior researcher, in this case a mathematics student: some actors generate conjectures, and others try to mathematically formulate and prove them[24].

It should be noted that even simple tasks such as: "Investigate the graphical changes in quadratic functions $y = ax^2 + bx + c$ when parameters

---

[23] This author refers to landscapes of investigation as being learning environments in mathematics classrooms, in which students are invited "to be involved in processes of exploration and explanation" (p. 67) with references to pure mathematics, to semi-reality or to real life situations.

[24] In this case, it should also be noted that one can think of an asynchronous collective, as Ricardo did not interact directly with the biology students, but through videotapes and their coursework. We are still analyzing the relationship between this kind of interaction and those that include the Internet and its synchronous and asynchronous relationships (see Borba, 2004, Gracias 2003, and Borba and Penteado, 2001).

a, b and *c* change", can turn into interesting problems for some students and can offer the teacher a chance to explore mathematics taking students' conjectures as a starting point, much like in the modeling approach. There are several other good examples discussed in GPIMEM papers, and others are still under analysis. It should also be noted that the graphing calculators were important actors in the collective that generated the conjecture, while paper and pencil were relevant actors at the moment of proving the conjecture.

The open-ended parabola task created an environment of investigation for the students. It is an example of the kind of activity proposed in the experimental-with-technology pedagogical approach. Even with more rigid tasks, very interesting mathematical conjectures have emerged. This is the case in our next example, regarding the teaching of conic sections.

3.   EXPERIMENTING WITH CONIC SECTIONS: MORE VISUAL CONJECTURES

The following example is based on some episodes that occurred during a teaching experiment using the software *Derive 5* for Windows, with a group of six volunteers, all women between the ages of 18 and 20, who were studying at the University of La Pampa - Argentina (Etcheverry, Evangelista, Reid, Torroba and Villarreal, in press). The students were enrolled in a first-year calculus course for pre-service mathematics teachers.

The pedagogical proposal included the study of graphs of some conic sections, analyzing changes in the graphical representations resulting from variation of the parameters in the algebraic expression $Ax^2 + By^2 + Cx + Dy + E = 0$. The students had already studied conic sections in their regular calculus course within a traditional approach, deducing their algebraic expressions from the geometrical properties that define them.

One of the activities that the researchers designed was for the students to study the effects of the variation of parameter C on the graph of equation $3x^2 + 3y^2 + Cx + 4y + 5 = 0$. Using a computer command that makes it possible to plot various graphs simultaneously, one of the students, named Ana, varied parameter C. She assigned integer values from -10 to 10 to C, generating, in this way, 21 equations. Only eight curves appeared on the computer screen (see Figure 7-3).

The student realized some graphs were not displayed. She decided to obtain the standard form ($(x - a)^2 + (y - b)^2 = R^2$) of the equation of each circle to find the coordinates of their centers, (a,b), and the values of their radiuses, *R*. In this way, she could find out if some values of *C* were failing to generate a circle. Ana presented their results in a table (see Table 7-1).

As she analyzed the table, she realized that the second column (R^2) had negative values, when *C* varied between -6 and 6, which implied that there were no circles for those values of *C*, since R^2 must be positive. Then she concluded that the graphs exist for integer values of *C*, from -10 to -7 and from 7 to 10. In this case, the student used algebra to explain the absence of some graphs on the computer screen.

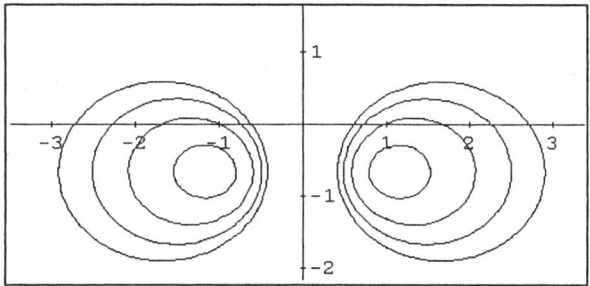

Figure 7-3. Graphs of circles $3x^2 + 3y^2 + Cx + 4y + 5 = 0$, assigning integer values from -10 to 10 to parameter C.

Table 7-1. Ana's table showing the coordinates of the center and radius for each *C*.

C	R^2	R	Center
10	1,55	1,24	$\left(-\dfrac{15}{9}; -\dfrac{2}{3}\right)$
9	1,03	1,01	$\left(-\dfrac{3}{2}; -\dfrac{2}{3}\right)$
8	$\dfrac{5}{9}$	0,74	$\left(-\dfrac{4}{3}; -\dfrac{2}{3}\right)$
7	0,139	0,37	$\left(-\dfrac{7}{6}; -\dfrac{2}{3}\right)$
6	$-\dfrac{2}{9}$	Does not exist	
5	-0,53	Does not exist	
.	.	.	.
.	.	.	.

Before we continue, it should be noted that Ana had found a discrepancy between the algebraic expressions she had typed in the computer and their graphical representations displayed by *Derive*. She then worked with a different algebraic approach, using paper-and-pencil, to generate a table so that she could investigate why some graphs were not being displayed on the computer screen. The coordination of the results found in different representations led to results that could later be shared with other students.

In the same teaching experiment, another student, Florencia, decided to assign integer values to C from -50 to 50 (but at intervals of 10). She observed the graphs on the computer screen; then she magnified the image using the *zoom* command (see Figure 7-4) and arrived at the following conclusion:

> For large values of C, [the circle] will never touch the axis of ordinates; it gets closer and closer to the axis, but never touches it.

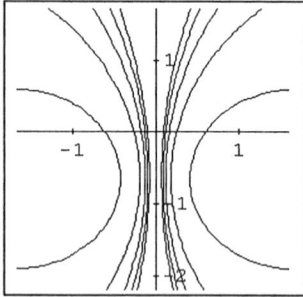

Figure 7-4. Graphs of circles $3x^2 + 3y^2 + Cx + 4y + 5 = 0$, assigning to parameter C integer values from -50 to 50 at intervals of 10.

This conjecture is true, but it was not proved at that moment[25]. The researcher proposed changing the parameters of the quadratic terms, but

---

[25] This conjecture led us to think of a possible proof. We would like to outline our reasoning. First of all, we obtained the standard form of the equation $3x^2 + 3y^2 + Cx + 4y + 5 = 0$ that is $(x + C/6)^2 + (y + 2/3)^2 = C^2/36 - 11/9$ showing that the center of the circle for each C is $(-C/6, -2/3)$. Then, we observed that these circles never cross the y-axis, since if $x=0$, the equation has no real solution. Then, we continued considering two cases: $C > 0$ and $C < 0$. If $C > 0$, the circles are in quadrants II and III, and the point on each circle that is closest to the y-axis is of the form $(x, -2/3)$ with $x(C) = \left(-C + \sqrt{C^2 - 44}\right)/6$. This function is increasing when $C > 0$, and $\lim_{x \to 0} x(C) = 0$ and, since circles never cross the y-axis, we have proved the conjecture. We could do an analogous reasoning for $C < 0$.

maintaining them equal, to observe the behavior of the graphs. Then Nancy worked with the equations $4x^2 + 4y^2 + Cx + 4y + 5 = 0$, $x^2 + y^2 + Cx + 4y + 5 = 0$ and $0.9x^2 + 0.9y^2 + Cx + 4y + 5 = 0$, and concluded that in every case, the circles did not intersect the ordinate axis. Meanwhile, Laura had been working with the equation $0.2x^2 + 0.2y^2 + Cx + 4y + 5 = 0$, and said that these circles crossed the y-axis at two points, as shown in Figure 7-5.

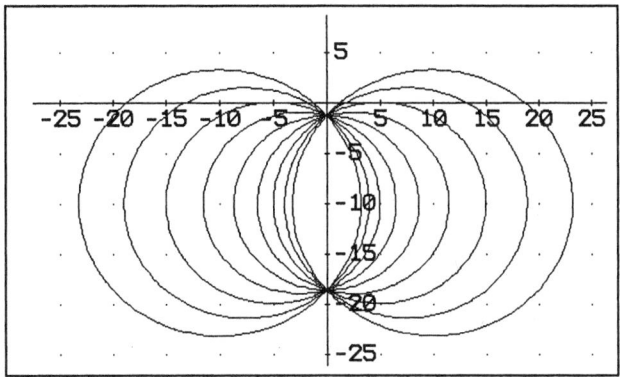

Figure 7-5. Graphs of circles $0.2x^2 + 0.2y^2 + Cx + 4y + 5 = 0$.

At that point, Laura proposed a new challenge: to find a value for the parameters of the quadratic terms such that the circles would be tangent to the ordinate axis. Having in mind their previous explorations, the students decided to vary parameter *A*, in equation $Ax^2 + Ay^2 + Cx + 4y + 5 = 0$, with values between 0.2 and 0.9. When they tried *A* = 0.8, they got graphs that appeared to be tangent to the y-axis (see Figure 7-6).

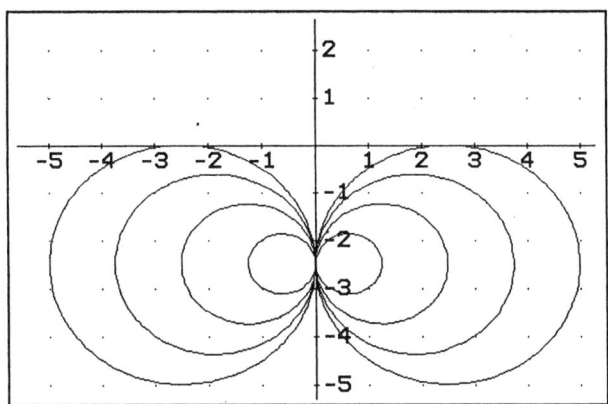

Figure 7-6. Graphs of circles $0.8x^2 + 0.8y^2 + Cx + 4y + 5 = 0$

They were amazed by this result. When the teacher asked if they were really convinced about that, Laura answered:

> I used the zoom to be sure that the circumferences were tangent. But I can solve the equations to find the tangent point.

Then, algebraic deductions were made to justify the conjecture. One of the students, Florencia, worked out the deduction of the result with the help of the teacher. She said she was looking for a point on the y-axis such that the circles cross the axis at that single point. She realized that the point would be of the form $(0, y)$, and evaluating the second-degree equation $Ax^2 + Ay^2 + Cx + 4y + 5 = 0$, at $x = 0$, she obtained $Ay^2 + 4y + 5 = 0$. At that point, the student again asked the teacher for help. The teacher suggested she solve the quadratic equation, and the student wrote the solutions $y = \left(-4 \pm \sqrt{4^2 - 4A5}\right)/2A$. Then, the teacher asked about the meaning of the two solutions she had obtained. This last intervention compelled the student to impose the value zero for the discriminant, since she wanted to get just one solution (the point where the circle is tangent to the y-axis). In this way, she finally solved the equation $4^2 - 4A5 = 0$, getting the expected value for $A = 0.8$.

This example shows, once more, that even with a restricted task such as "Study the graphs of equations $3x^2 + 3y^2 + Cx + 4y + 5 = 0$, varying the values of C", the collective of students-with-computers became engaged in the generation of various interesting conjectures, and new challenges were posed by the students themselves, as when Laura proposed finding circles tangent to the y-axis.

This example also illustrates how a collective of students-with-computers engaged in mathematical experimentation and generated several conjectures associated with the visual feedback of the computer. The conjectures were neither inside the students' minds nor outside of them, on the computer screen, but rather were generated together with the computers; and even if a particular conjecture is attributed to a single student, the collective, as a whole, conditioned its emergence. The 'experimentation approach in action' illustrates how the coordination of multiple representations can be a path for the generation of knowledge in the classroom.

The immense quantity of graphs that can be plotted using the software, without having to input expressions one by one, when adequate commands are used, transforms the computer into a laboratory for mathematical experimentation. The students also used the *zoom* command as a tool to verify the validity of their conjectures. The *zoom* is a computational resource they incorporated as a frequent strategy to ensure the validity of their conjectures. They believed in graphical, i.e. visual, verification. The algebraic deductions were done because the teacher strongly encouraged them to do so.

The visual and experimental possibilities of collectives in which information technology is an important actor provide the students with means to verify the conjectures they generate. They also make it possible for teachers to suggest the need for proof in an environment where the students are already attempting to coordinate different representations such as graphs, tables and algebraic expressions. Teachers can then verify how far they want to go with their 'proof agenda', depending on their audience. They can also have multiple agendas, assigning the algebraic proof as a task for some students, visual proofs for others, and 'empirical' arguments for still others. Comparing these approaches in a non-hierarchical way can extend the notion of multiple representations in directions other than those developed by Borba and Scheffer (2003, in press). In this case, we want to suggest that different kinds of arguments can be coordinated, in the sense that students who have coordinated multiple representations are coordinating arguments more-or-less linked to those representations.

4. EXPERIMENTING WITH FUNCTIONS I: THE AG-GA THEOREM

The *zoom* resource, and the use of specific cases to verify conjectures, are common strategies at various levels of education, as can be seen in our next example, which involves students from the first year of high school. This example comes from teaching experiments conducted by Benedetti (2003) with a pair of students, named Gianluca (G) and André (A), using a shareware version of *Graphmatica*, which is available on the Web. In the episode we present, the students were talking about the graphs of $y = x^2$ and $y = x^3$, which they had just plotted in the computer (Figure 7-7).

They observed that both graphs pass through the point (1,1) and Gianluca asserted that the value of a in both equations is 1, referring to the coefficient a in polynomial functions $y = ax^n$ with $n \in N$. This fact inspired him to guess that if coefficient a were equal to 2, the graphs of $y = 2x^2$ and $y = 2x^3$ would pass through point (2,2). After doing the graph on the computer, the students realized that the interception point was (1,2), not (2,2). Looking at the graphs, Gianluca observed that "*a is on the y-axis*" (see Figure 7-8), meaning that the ordinate of the point where both pairs of functions intersected each other coincided with the value of a.

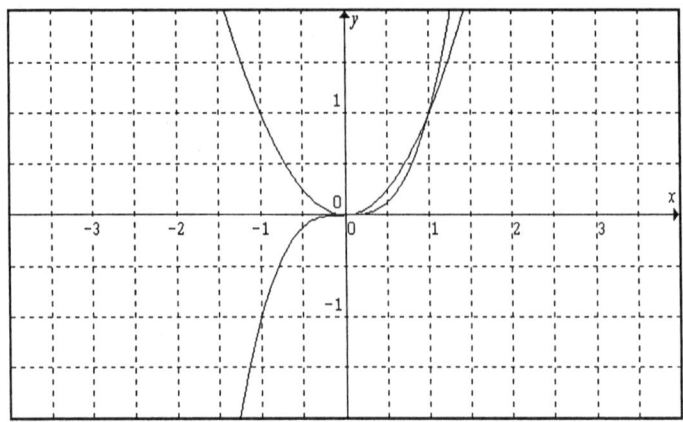

Figure 7-7. Graphs of $y = x^2$ and $y = x^3$.

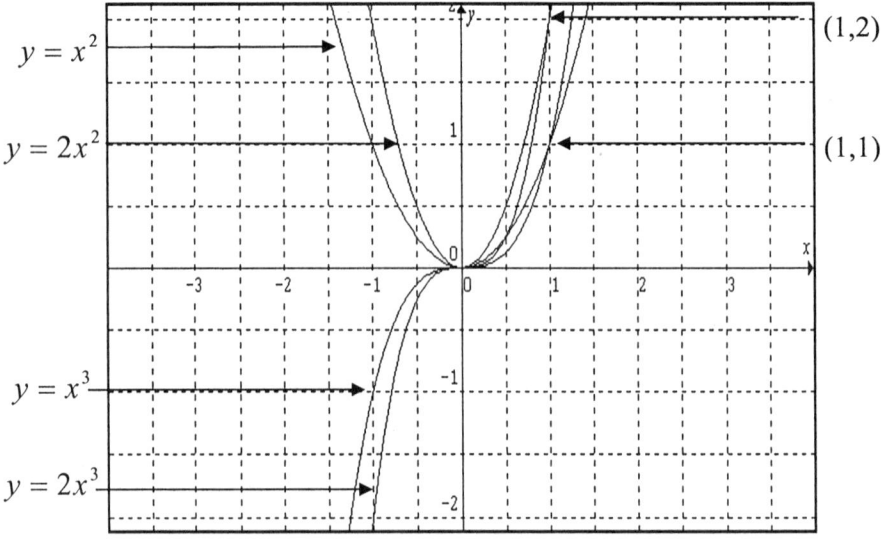

Figure 7-8. The first pair of functions is $y = x^2$ and $y = x^3$. The second pair of functions is $y = 2x^2$ and $y = 2x^3$. For each pair of functions, their intersections at points (1,1) and (1,2) are marked.

That observation, and the experimentation with different functions, lead the students through an 'educated' trial and error process that would facilitate the generation of a conjecture that we could state as follows: functions with the formula $y = ax^n$ pass through point $(1, a)$. The interviewer (I) asked the students to prove that $y = 2x^2$ and $y = 2x^3$ intersected at (1,2), and the following dialogue ensued:

I: How do I know that the two graphs intersect at the point (1,2)?

A: Ah, by the drawing! Look at 1 and 2 [showing the coordinates on the computer screen]

I: By the drawing?

A: Yeh, by the drawing.

The interviewer wanted the students to verify algebraically that the point (1,2) belonged to the graphs of both functions, but a visual verification was sufficient for the students at that moment. Nonetheless, the interviewer insisted, and Gianluca used the computer to verify that points (1,1), (1,2) and (1,3) belonged to the pairs of functions $y = x^2$ and $y = x^3$, $y = 2x^2$ and $y = 2x^3$, and $y = 3x^2$ and $y = 3x^3$, respectively. Gianluca observed that the functions that pass through point were 'specific', meaning that they were of the form $y = ax^n$. To be sure of this fact, he plotted the graphs of $y = 4x^2 + 3x - 12$ and $y = 4x^2 + x$ to verify that, if the values of the coefficients of the non-quadratic terms were not zero, the functions may not necessarily pass through $(1, a)$, in these particular cases (1,4). The student also plotted graphs of functions $y = x^5$, $y = x^6$, $y = x^7$, varying the exponents of x, and verifying that all of them passed through point $(1, a)$, with $a=1$ in these cases. After these graphical verifications, the interviewer insisted:

I: So, our problem now, kids, is the following: prove to me that all functions of the type ax^3, ax^5, ax^6 ... all of these, pass through the point $(1, a)$.

The students engaged in that task, even though they themselves considered the graphical verifications to be sufficient. They started working algebraically with paper and pencil, verifying that the point (1,2) belongs to the graph of $y = 2x^2$, as shown in Figure 7-9

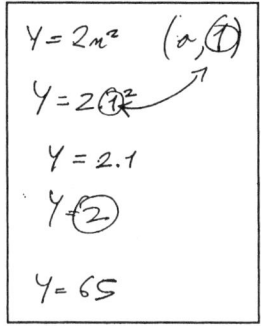

Figure 7-9. Students' annotations. Observe that the order in the mathematical notation of the ordered pair is changed.

Then they went a step further when André decided to change the value of coefficient *a* in the function $y = ax^n$

A: Hey – I think I found something! [...] No, but just look. It's really easy. If we take, for example, the . . . lets put any number in the place of a, right? If we put . . . 65, right?, if we put one in the place of *x* [referring to x^2], it'll give a!

G: It'll give 65. It'll give (1,65).

A: It'll give 65. [writing $y = 65$ in the last line of Figure 7-9]

Finally, the students, with the aid of the interviewer, arrived at a generalization that André expressed as follows:

A: It doesn't matter what power I raise it [referring to *x*] to, it doesn't matter... what number I put in front of it [referring to the coefficient of *a*], it will always give *a*; no, wait . . . I mean, 1 and *a* . . . no, *a* and 1? [looking at G].

André was not sure about the order of the coordinates of the ordered pair: $(1,a)$ or $(a,1)$? After a brief discussion, they produced a written text. The students and the teacher assigned it the status of 'theorem', and called it AG-GA, using the first letters of the students' names. They stated it in the following way:

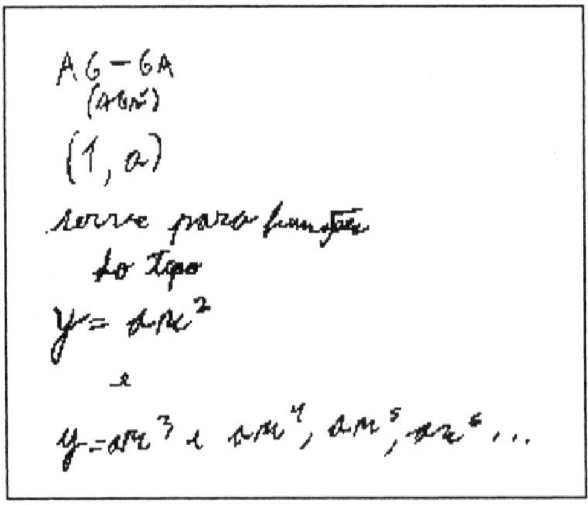

$AG - GA$ *(AGs')* $(1, a)$ *serve para funtar* *to tipo* $y = ax^2$ $y = ax^3$ *e* ax^4, ax^5, ax^6, . . .	AG-GA (AGs') $(1,a)$ *works for functions* *of the type* $y = ax^2$ *and* $y = ax^3, ax^4, ax^5, ax^5, ax^6...$

Figure 7-10. The AG-GA theorem in its original version and its translation into English.

It is worth noting that polynomial functions $y = ax^n$ with $n>2$ were new for the students; they had never seen their graphs before. André and Gianluca explored them graphically using the software; they were able to talk about them fluently, making comparisons and formulating a conjecture, whose validity they tested using particular functions. It was apparent that the students made algebraic verifications only because the interviewer required them to do so. For the students, it was an unnecessary step, because they had been convinced visually by the graphs on the computer screen that resulted from their experimentation with different functions. It is interesting to note that Gianluca established restrictions for their conjecture: it was valid only for specific functions of the form $y = ax^n$, which was a necessary condition to his conjecture. He showed examples of functions that do not verify the conjecture because they were not of that form. He presented counter-examples to convince himself, his colleague and the interviewer. He was verifying the necessity of the condition he had established, an important mathematical activity in the process of generating and validating, or refuting, a conjecture.

This example, and the two previous ones, regarding quadratic functions and circles, show how different mathematical conjectures were generated visually. Some of them were proved and others were accepted without any proof. The computers or graphing calculators were the media with which the students created and 'visually proved' their guesses. The question seems to be: for whom (what audience) and in which cases (what particular tasks) is a mathematical proof necessary? We think the previous examples enable us to think about the place of proofs in an experimental-with-technology approach, as we discussed previously when we referred to proofs in modeling projects in Chapter 6. In a very interesting essay, Thurston (1994) reflects about proof in mathematics and, speaking from his own experience as a mathematician, he says: "It becomes dramatically clear how much proofs depend on the audience. We prove things in a social context and address them to a certain audience" (p. 175). He goes on to say that a proof that could be communicated in two minutes to topologists would need an hour lecture before analysts would begin to understand it (or vice versa). Thurston was talking about an audience of mathematicians, but his essay somehow inspired us: in the particular learning contexts of our students, they created particular criteria of validity and truth that may include graphical argumentation or computer commands as ways to prove their conjectures. That doesn't mean that the teacher, as a member of the thinking collective, can't draw attention towards mathematical proofs, as was the case with Renata's conjecture, when the teacher presented a proof of the conjecture using the mathematical tools the students had just learned; the case of Ricardo, the math student who proved the conjecture of a biology student; and the case of Gianluca and André, who engaged in an algebraic process to prove their statement.

In all the examples we have discussed thus far, algebraic representations of functions or circles were the starting point to initiate the exploration, with the graphing calculator or with the computer. In the next section, we would like to present an example where the algebra step was bypassed using a different technological interface.

5.   EXPERIMENTING WITH FUNCTIONS II: MULTIPLE REPRESENTATIONS AND INTERMEDIA COORDINATION

This example comes from research conducted by Scheffer (2001, 2002) and Borba & Scheffer (2003, in press), who analyzed the mathematical activities performed by 8th grade students using CBR (Calculator Based Range)[26], a motion detector (sensor) linked to a graphing calculator that provides graphs of distance × time while movements are executed (see Figure 7-11). For instance, a student walking towards a wall at a constant speed, with the sensor pointing at it, would generate a decreasing linear function, since the distance between the sensor and the wall decreases as (real) time passes.

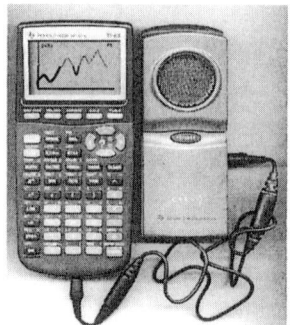 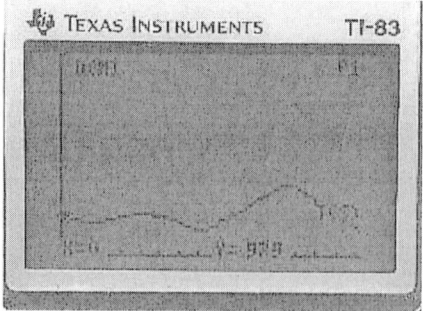

Figure 7-11. On the left, a graphing calculator with a CBR. On the right, the calculator screen showing a graph generated with the data collected by CBR.

A teaching experiment was conducted with Rafael (R), a 15-year-old boy, and Queila (Q), a 14-year-old girl, both of whom had been introduced to the calculators previously in their classroom. The students, in a research lab environment, were asked to make any movement they wanted to with the sensor, and then guess what graph (distance × time) the calculator would present as they moved the CBR together with their own body. The main goal

---

[26] This equipment was developed by Texas Instruments (www.ti.com). We thank T.I. for the support provided for this part of the research of GPIMEM.

of this open-ended problem was to connect a 'free' body movement, performed by one of the students, to the Cartesian graphs generated by the motion detector on the graphing calculator screen. Rafael took the lead with the CBR. He held it close to his body and walked towards the wall with the CBR pointed at the wall. As he drew close to the wall, he stopped briefly and walked backwards with the CBR still pointed towards the wall. Rafael drew a graph like the one in Figure 7-12.

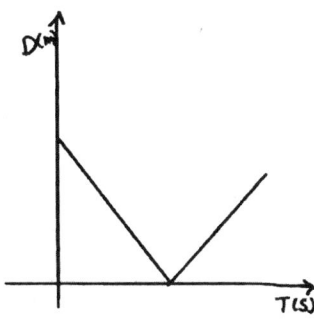

Figure 7-12. Rafael's graph.

As seen in Figure 7-12, Rafael drew a V-shaped graph touching the x-axis. The researcher (Ni)[27] presented another probe and brought Queila into the debate

Ni: Did it touch the wall?

R: Right, almost, it seems like I got really close, no, and then I did... like this, I think. [referring to the graph on the blackboard]

Ni: Uh-huh. Do you agree with that? [talking to Queila]

Q: Well, according to what he said, I agree.

Ni: Hmm. So, you got close- you started far away, moved closer, and then got there and came back.

R: Yeh, I came back to the place I had started out from.

Q: I would only mark a pause there, right? . . because he stopped there.

Ni: Ahhh! It's missing a pause! Go to the board and draw it like you think it would be with a pause ... [inaudible] Do it with another color; get the orange chalk.

Q: So how should I do it now?

Ni: Do it on top of the same graph.

---

[27] We use this abbreviation for the researcher to avoid confusion with our references to Rafael (R). The interviews were conducted by Nilce Scheffer.

The researcher suggested that she sketch the new graph in the same Cartesian frame of the first graph, so Queila drew a line corresponding to the dotted line in Figure 7-13.

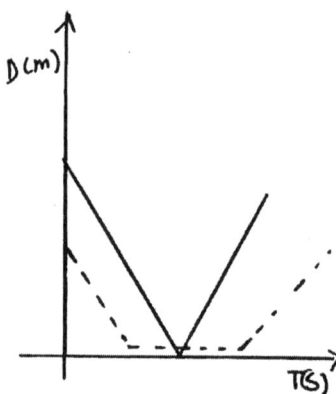

Figure 7-13. The dotted line corresponds to Queila's graph.

The dotted line Queila drew seemed to solve the problem in a way that pleased all three participants, representing Rafael's brief stop with a line parallel to the x-axis. The researcher then showed them the graph generated by the calculator as Rafael walked back and forth towards the wall (Figure 7-14).

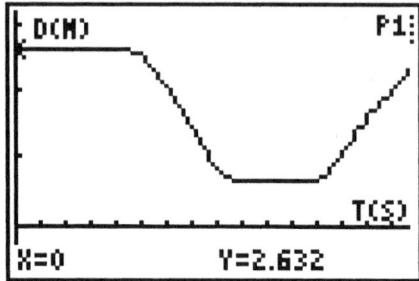

Figure 7-14. The graph on the calculator screen.

Ni: So, the graph on the calculator came out like this [showing them the calculator. See Figure 7-14].

R: It came out really different from the one on the board [referring to the V-shaped graph in Figure 7-13].

Q: It didn't come out so different [referring to the dotted line in Figure 7-13, drawn by Queila].

R: No, hers came out the same [also referring to Queila's dotted line].

Ni: But you understood the idea, didn't you, Rafael? Because you made, ah . . you moved closer and then moved away.

Q: He moved, right?

Ni: Uh-hmm.

Q: So, there was the movement he made, that he moved closer, right?

Ni: Right.

Q: He got there and stopped. Then I think it should have this pause, because afterwards he came back, right?

Ni: Hmmm.

Q: He was moving away. Then he got there, moved closer, there was a pause, when there was a line, and he went back. He moved away, right? I think it would turn out like this.

Rafael was surprised by the fact that the graph drawn by the calculator was more similar to Queila's than to his own, and at that moment he realized the qualitative difference between both graphs drawn on the blackboard. It is important to note that, even though the graph of distance × time of the movement appeared simultaneously on the graphing calculator screen as it was performed, the researcher decided to show it to the students only after they had made their guesses regarding the graph to stimulate the discussion. The interviewer and the students analyzed the graph on the calculator screen and compared it with the graphs on the blackboard. The interviewer called their attention to the first constant piece of the graph, pointing out that the CBR was functioning before Rafael started moving. Then she reproduced the graph created by the calculator on the blackboard, and Queila drew arrows 1, 2 and 3 (see Figure 7-15) to indicate the parts where Rafael had moved toward the wall, stopped, and moved away, respectively.

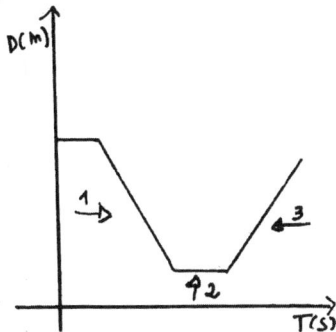

Figure 7-15. The graph made by the interviewer and the arrows drawn by Queila.

This episode, which ended here because there were no more discrepancies between the chalk-and-blackboard and the graphing calculator representations, illustrates a particular way of coordinating representations. Students first had to come to an agreement, with the help of the interviewer, regarding which graph, drawn on the blackboard, would best represent the movement Rafael performed. Secondly, they had to coordinate the experience of the body movement with the graph drawn on the board and the one generated by the calculator as Rafael walked. The episode emphasizes how body movement was coordinated with established mathematical representations. It should be noted that the point where he paused was a point of difficulty in terms of Cartesian graphical representation. This may be due to the fact that, at this point, there was a greater discrepancy between the body movement and the Cartesian graph. The student experienced the pause physically as a lack of movement; he stopped moving. But at the same time, he could see that the line on the Cartesian graph did not 'pause' when he did. Coordinating the physical experience of being stopped with the dynamic, continuous line on the Cartesian graph that appears on the calculator screen is not a trivial task.

Much of the research that has been developed regarding mathematical representations, and the coordination of multiple representations specifically, usually refers to numerical, algebraic and graphical representations. The discussion about multiple representations acquires a new dimension when representations are associated with the media that produce them (oral, written or computational media), or when body awareness is taken into consideration. This perspective brings up the need for intermedia coordination, and broadens the possibilities of visualization, since kinesthetic images[28] (Presmeg, 1986b) belong to the repertory of possible images related to mathematical concepts.

In addition to the coordination of representations produced by different media, there are two more aspects of the episode with Queila and Rafael that should be noted. The first is that the episode serves to inform the discussion regarding the dichotomy between visualization as an internal, mental process and visualization as an external process using some medium. When the student performed the movement together with the CBR, it became 'part of his body'; it became one with him - not just because he was holding the CBR at one point, but also because, once the student used the interface, he incorporated that experience and reflected on it. From this perspective, an internal/external dichotomy makes no sense, and it becomes clear that we are dealing with a collective formed of the student and the CBR. Secondly, the episode presents a visual approach to the introduction of functions that

---

[28] Images involving muscular activity, for example, gestures.

bypasses algebra, at first, showing that the generation of functions does not depend on the pre-existence of an algebraic expression. The generation of piecewise-defined functions emerges more naturally when one looks at the graph produced by the graphing calculator, and the discussion regarding how to generate an algebraic expression for the graph follows naturally, as well.

6. CONSTRUCTION OF DERIVATIVES: A GRAPHICAL APPROACH

This example comes from research developed by Villarreal (1999, 2000), who conducted teaching experiments with first-year biology major students, aged 18 to 21, enrolled in the 1997 class of the applied mathematics course mentioned previously. The students who volunteered to participate in the research project had little experience with computers or mathematical software. The aim of the research was to describe the mathematical thinking processes of the students as they worked with mathematical questions using a DOS version of software *Derive*[29].

Camila is the protagonist of this particular episode. The problem posed to her was to sketch the graph of the derivative of a function without knowing its algebraic expression. Camila had worked previously with this kind of problem in an earlier session of the teaching experiment. On that occasion, she had established a relationship between the growth of the function and the sign of its derivative:

In the part where the function is decreasing, the derivative will be negative; in the part where the function is increasing, the derivative will be positive.

Camila had learned in her mathematics course that the derivative of a function at a given point x_0 was the value of the slope of the line tangent to the curve (if it exists) at point $(x_0, f(x_0))$. She knew that the differentiable extremes of the function determine the roots of the derivative, because the slopes of the tangent lines at those points are zero, and she used that information to sketch the graph of the derivative. After showing the graph of F on the computer[30], a transparency was taped over the screen and Camila drew her graph of F' on it (Figure 7-16)

---

[29] New versions of *Derive*, like *Derive for Windows*, are now very popular, but at the time of the teaching experiments (1997) a *Derive for DOS* was the only version we had access to.

[30] In case the reader wants to reproduce the function $y = F(x)$ in the computer, the algebraic expression of it is: $y = x(x^2 - 1)(x - 2)^2$

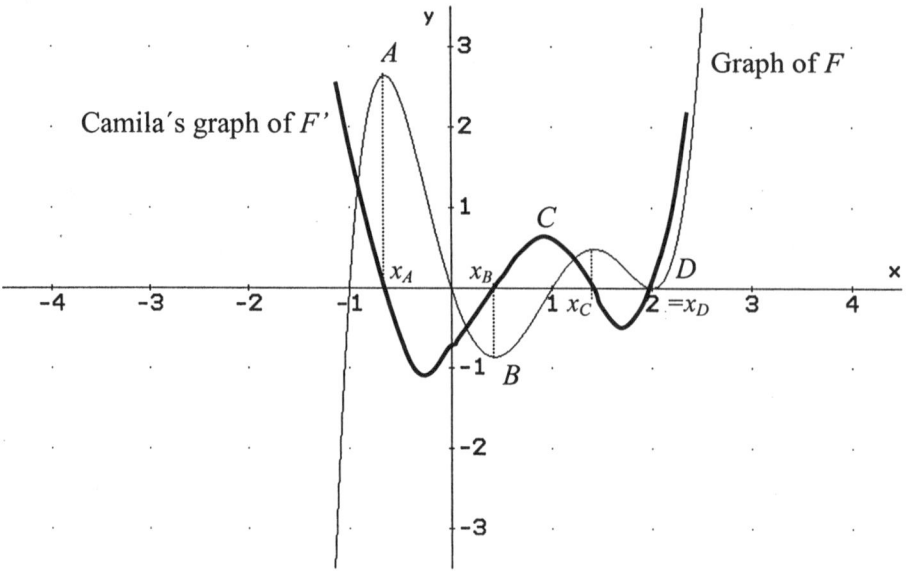

Figure 7-16. Camila's graph of F'. The labels x_A, x_B, x_C, x_D, A, B, C, D were not in the
original figure, but we include them here to facilitate future references to these points.

At this point, Camila wondered how much the derivative was *"going up
or down"* in the intervals between their roots, i.e. where the extremes of the
derivative were. The interviewer asked: *What would it depend on? Do you
have any idea?,* and she answered:

> On the slope of this part, right? [indicating the part between *A* and *B*], ...
> since this part here [indicating between *A* and *B*] decreases more than this
> part here [indicating between *C* and *D*], then this part here [indicating the
> derivative between x_A and x_B] would come down further than this part
> here [indicating the derivative between x_C and x_D].

It seems to us that Camila was suggesting that a way to find out *"how
much the derivatives goes up or down"* could be by approximating the
function with a straight line in the intervals $x_A < x < x_B$ and $x_C < x < x_D$,
since she said the *"slope in this part, right?"*. This could be considered a
visual and qualitative criterion to decide about *"how much the derivative
goes down"*.

The interviewer showed the graph of the derivative on the computer
screen so the student could compare it with the one she had sketched.

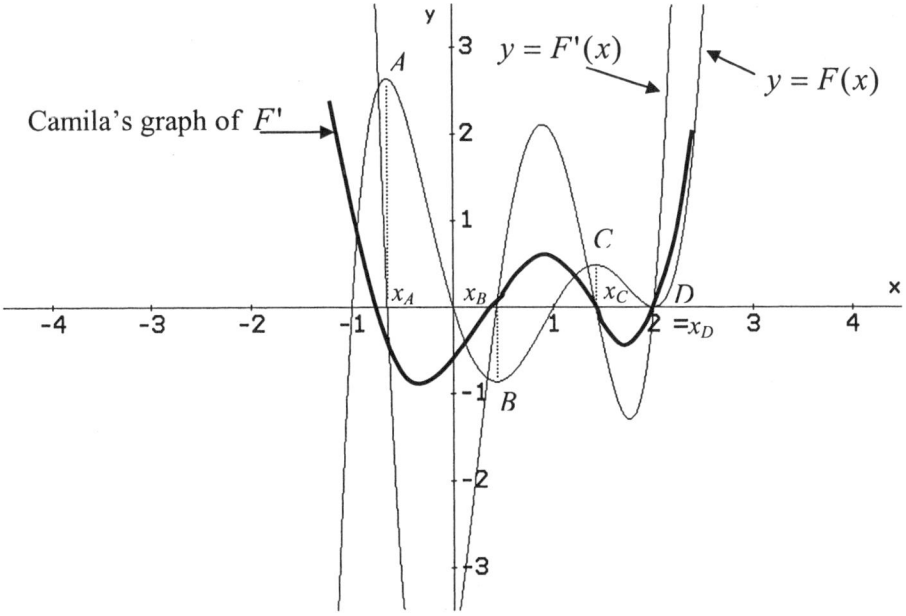

Figure 7-17. Camila's graph of *F'* and the graph provided by the computer.

Camila observed that, in the interval $x_A < x < x_B$, the graph of the derivative "*is a lot lower*" and wondered why. She continued associating it with the "*slope of the graph*" with respect to the y-axis, between the extremes of the function *F*. For example, if the graph of the function is decreasing and more sloped towards the y-axis, then the derivative will reach smaller values, that is why the derivative attains smaller values in interval $[x_A, x_B]$ than in interval $[x_C, x_D]$. The comparison of her graph and the graph plotted on the computer reinforces Camila's approach. She created a local visual criterion to answer her question, which we interpreted as referring to the extremes of the derivative.

The interviewer didn't know exactly how to take Camila's local criterion further, and suggested that she should consider lines tangent to the curve at each point of it. The student analyzed the derivative's behavior using tangent lines. Camila knew that the slopes of the lines tangent to the function at their extremes were zero, and that was why the derivative would be zero at those points, and its graph would therefore cross the x-axis. Now she had to apply this process to other points of the function.

The interviewer proposed that she trace with a ruler, on the transparency taped to the computer screen, the line tangent to the curve at point $(0.75, -0.5156)$. Camila traced the line as shown in Figure 7-18[31].

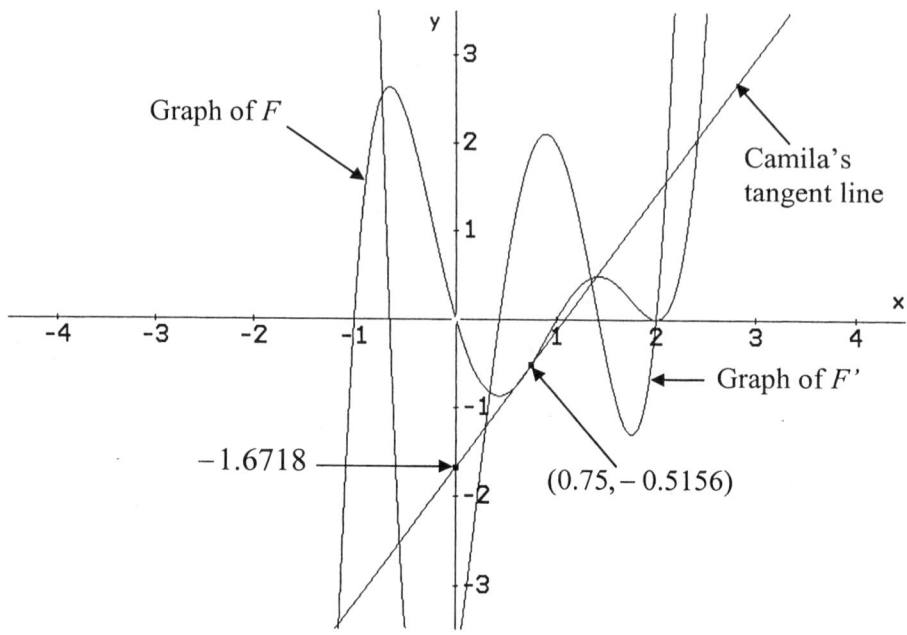

Figure 7-18. Camila's line tangent to the curve at point $(0.75, -0.5156)$.

Then, Camila calculated the slope (a) of the line, with the equation $y = ax + b$, where (x,y) was the point where the line was tangent to the curve (in this case $(0.75, -0.5156)$), and $b = -1.6718$, the y-intercept of the line Camila drew, was determined using the computer cursor. The student solved the equation:

$$-0.5156 = a\,0.75 + (-1.6718)$$

and obtained the value of the slope of the tangent line, $a = 1.5416$.

The interviewer asked her to interpret the information graphically. Initially, Camila had some difficulty interpreting the point $(0.75, 1.5416)$ as being a point on the graph of the derivative, but since she could see the graph of the derivative on the computer screen, she finally indicated that the

---

[31] The interviewer selected this paricular point using the trace command that makes the cursor move throughout the curve while showing the coordinates of the point where it is.

point $(0.75, 1.5416)$ belonged to the graph of the derivative. The computer showed that the derivative actually passed through the point $(0.75, 1.8906)$ and not through $(0.75, 1.5416)$, as her calculations had indicated. The interviewer proposed to draw the tangent line using a computer command. Camila was able to compare the line she had sketched with the one the computer plotted, realizing that the line she had drawn on the transparency had a smaller slope than the one displayed by the computer.

In summary, this episode illustrates how a student sketched a graph for the derivative of a function without knowing its algebraic expression. She used previous knowledge to decide about the roots and signs of the derivatives that were associated with the extremes of the function and its increasing or decreasing intervals. After sketching her graph, she wondered about the extremes of the derivative. First, by making a linear approximation of the function between its extremes, she was able to decide that, in the interval $[x_A, x_B]$, the derivative would have smaller values than in interval $[x_C, x_D]$. Although this visual criterion cannot be considered a precise procedure to exactly determine the extremes of the derivative, it served as a criterion to decide in which of the intervals where the function was decreasing the derivative would attain smaller values. Secondly, the student learned to graphically construct the derivative function, point-by-point, associating to each x the value of the slope of the line tangent to the curve at point $(x, f(x))$.

Camila had also sketched the derivative of other functions, using the strategy of drawing tangent lines at some points of the function and calculating their slopes, to determine different values of the derivative. For example, to sketch the graph of the derivative of the function in Figure 7-19, she used the computer to decide the value of the derivative at point $x = -1$. She knew that the function was decreasing in interval $(-\infty, A)$, and that the derivative would therefore be negative there, but she hesitated about the tangent line at $x = -1$, where it seemed to be horizontal. The software played a relevant role in this particular humans-with-computer collective because, after drawing the line tangent to the function at $x = -1$ with the computer, she saw that its slope was zero, and that the derivative would then have another root at that point, apart from the roots A, B and C that she had previously determined to be points where the tangent lines were horizontal.

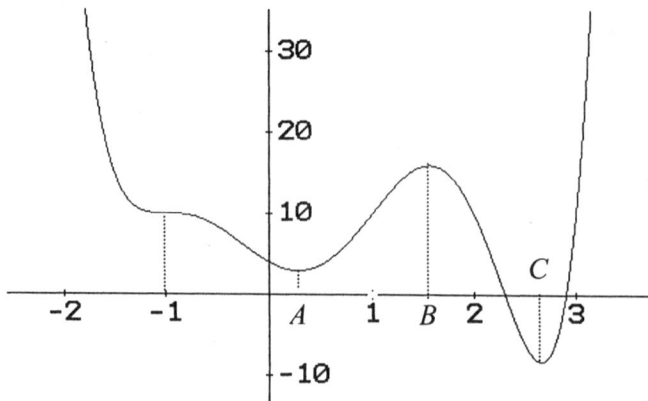

Figure 7-19. Another function given to Camila to sketch the graph of its derivative.

The important point of this example is the generation of a particular visual process to better understand the construction of the derivative. Camila had studied derivatives in the mathematics course she was enrolled in at the time of the teaching experiment, and her experience with the computer throughout the teaching experiment reorganized her knowledge about that topic, since she worked in the graphical construction of the derivative of a function without having its algebraic expression, compelling her to invent visual procedures and strategies with the aid of the computer resources. The student, together with the computer, observed that a, the value of the slope of the line tangent to the curve at (x_0, y_0), corresponded to the value of the derivative at x_0, and that (x_0, a) was therefore a point that belonged to the graph of the derivative.

The particular visual approach to constructing the derivative of a function that we have just described was generated in a teaching experiment with a particular student-computer-interviewer-slide-and-pen thinking collective. With paper-and-pencil, we can also sketch a qualitative graph of the derivative of a given function graph, trace tangent lines with a ruler, and try to measure their slopes. One difference between both processes is that the paper-and-pencil process is slower than the computer process, but there is yet another paramount difference: the feedback of the computer, and the facility of comparison with the graph of the derivative on the computer screen, which transform the computer into a 'conversation piece' (Meira, 1998), into a co-author of the student's ideas, helping her think about her answers, to correct them, to raise questions and to pose new answers.

7. TANGENT LINES: VISUAL AND ALGEBRAIC APPROACHES

This example, featuring Mayra and Carolina, also comes from the teaching experiments developed by Villarreal (1999, 2000) with first-year biology students using a DOS version of software *Derive*. These students were learning to use a software command that makes it possible to determine the equations of lines tangent to a given curve. The line tangent to $y = x^2$ at $x = 2$ was determined to be $y = 4x - 4$, and the graphs were produced (see Figure 7-20)

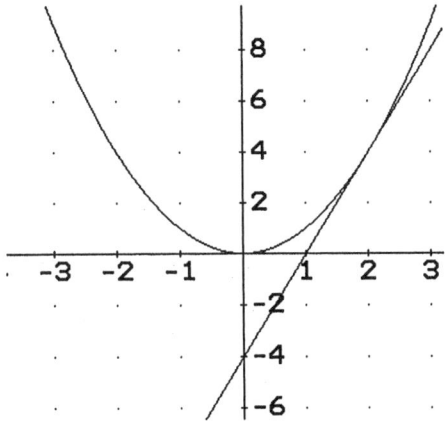

Figure 7-20. The graphs of $y = x^2$ and the line tangent to it at x = 2.

A conflict arose with the graphs shown on the computer screen, where the tangent line appeared to be touching the parabola at more than one point. This provoked the following question from Carolina: *"Can a tangent line touch at various points?"*

It is important to note that this conflict would not have arisen in a paper-and-pencil environment, since the condition of the straight line 'touching just one point of the curve' preceded the graph. On graphs drawn with paper and pencil, the points where tangent lines touch the graph are commonly represented with a single 'fat' dot, as shown in Figure 7-21, not unlike computer-generated graphs, but in the case of paper-and-pencil, the fact that the tangent line appears to be touching the curve at various points is simply disregarded.

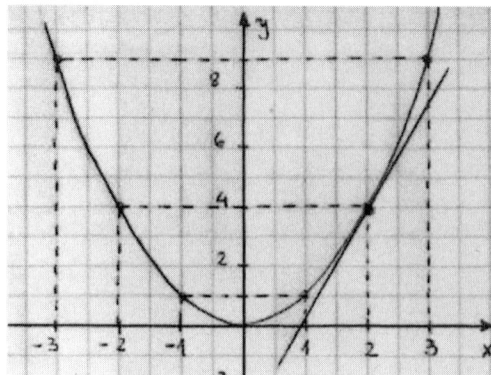

Figure 7-21. A paper-and-pencil graph of and the line tangent to it at *x*=2.

When the graphs are sketched with paper-and-pencil, there is no problem with the unity of the point where the curve and the straight line are tangent, because we control the representation. In the computer environment, however, the computer traces the graphs, and a discrepancy emerges. The students' first strategy to address the discrepancy was to use the *zoom* command to get a closer and better view of the region where the tangent line and the parabola appeared to be 'touching at various points'. However, this strategy failed to solve the problem, since the parabola and the tangent line tend to become indistinct in the neighborhood of the point where they are tangent (see Figure 7-22), which provoked Carolina to comment: "*It can't ever get a tangent line*".

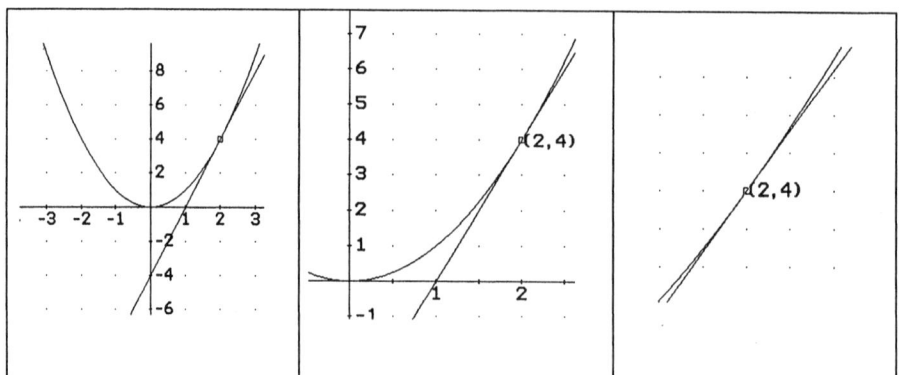

Figure 7-22. Successive *zoom in* in the neighborhood of (2,4).

The discrepancy between a graphical representation of a line tangent to the curve, and the students' concept of tangent line as being a line touching a curve at only one point, lead to the following "why question" (Borba, 1994): Why does it look as if there is more than one point where the curve and the line are tangent? Attempts to solve the discrepancy within the graphical representation, using the *zoom*, were not successful. The students finally decided that they should "*set the two equations equal to each other... the tangent line equation with the parabola equation*" and solve that equation to see if it had just one solution. The student's comment implied that she had an algebraic path to solve the discrepancy she found in the graphical representation. The algebraic facilities of *Derive* made it possible to find the equation of a particular tangent line using a special command of the software. The graphical representation of that tangent line led to a conflict, and then two strategies to resolve it arose: the first visual, and the second algebraic.

From the point of view of Aspinwall, Shaw and Presmeg (1997), who talk about uncontrollable images in mathematics, we could say that this episode shows that the graphical approach, encouraged by the computer, may generate uncontrollable images for the students, in the sense that it could introduce some conflicts or barriers to mathematical understanding. However, uncontrollable images may also appear in the algebraic approach. For example: the fact that a straight line $y = ax + b$ is tangent to a curve $y = f(x)$ does not mean that the line does not intersect the same curve at other points aside from the point where they are tangent. In this case, the equation $f(x) = ax + b$ would not have a single solution, as is the case of the lines tangent to parabolas at any point.

We can then say that algebra can reinforce conceptual understanding that, for example, in the case of a cubic function $y = x^3$, all lines tangent to it (except at $x = 0$) would also contain another point of the curve distinct from the one where the line is tangent to the curve. But what may not be easy for most students is to see algebraically that the system:

$$\begin{cases} y = x^3 \\ y = 3x - 2 \end{cases}$$

(in which the second equation is a line tangent to the curve represented by the first equation) has two distinct real solutions. As we plot the functions, using almost any software with graphing capabilities, it is easy to 'see' the existence of these two solutions (see Figure 7-23).

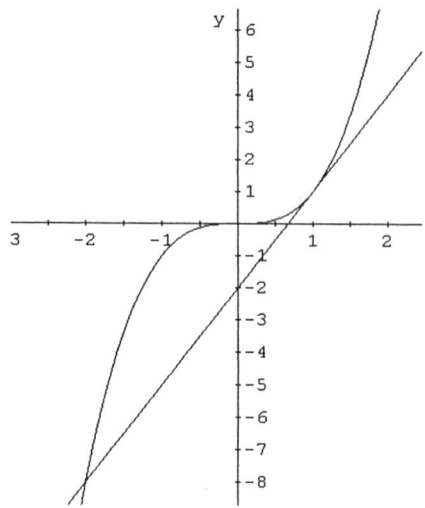

Figure 7-23. $y = x^3$ and a tangent line $y = 3x - 2$.

Moreover, we can also say that the paper and pencil media could lead us to "uncontrollable images"; for example, if one were to draw this tangent line ($y = 3x - 2$), one would more likely draw something like the drawing on Figure 7-24.

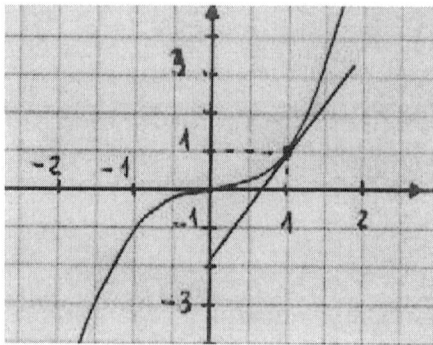

Figure 7-24. A paper-and-pencil sketch of $y = x^3$ and $y = 3x - 2$.

In our view, there is no 'good' or 'bad' medium. Media condition our thinking and the knowledge we produce. Media condition the representations we generate with them and offer different opportunities to understand mathematics. Since many different media are available, the model developed in Borba (1994) has to be extended to include intermedia coordination, as proposed by Villarreal (1999, 2000).

Let us return to our episode with Mayra and Carolina to discuss another point. The students' strategy of using the *zoom* command to verify that a line tangent to a parabola touched it at only one point proved unsuccessful, but rather than thinking that the graphs displayed by the computer could generate an incorrect image about the notion of tangency, one could think of the *zoom* strategy as a kind of visual demonstration of the fact that: the line $y = 4x - 4$ tangent to $y = x^2$ at (2,4) is the best local linear approximation of the parabola at that point. The local straightness of a function is a graphical way of verifying its differentiability at a given point. This computational visual strategy was used in teaching experiments conducted with two other biology students in order to verify that there is no possibility of having a straight line that locally approximates the curve shown in Figure 7-25 at point $x = A$, showing that the function is not differentiable at that point.

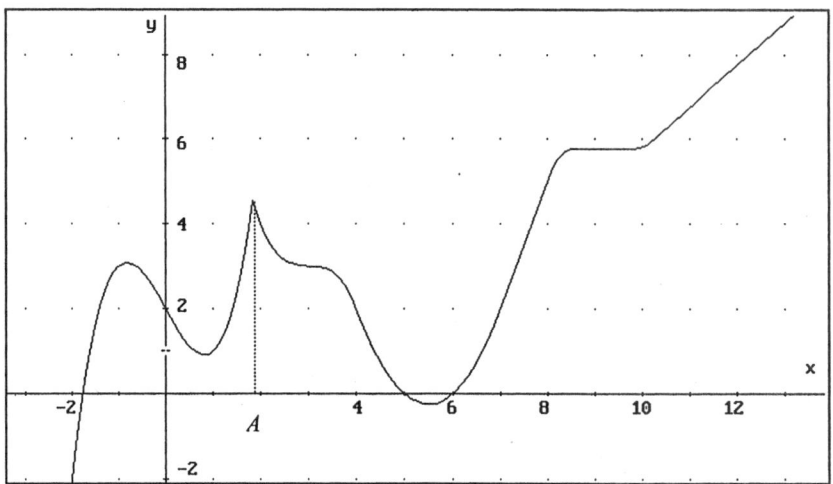

Figure 7-25. A function with a non-differentiable point at *x=A*.

The determination of non-differentiable points in a continuous function, such as the one in the previous Figure 7-25, may be easier using a visual approach like the *zoom* strategy, although we recognize that students may have some difficulties inputting piecewise-defined functions like the one displayed in Figure 7-25 because of the syntax used by the software. The *zoom* strategy is not new in the literature. Tall (1991) referred to a locally straight approach to calculus, and more recently Giraldo, Carvalho and Tall (2003) talked about it using the concept of *theoretical-computational conflict* ("any pedagogical situation with apparent contradiction between the

mathematical theory and a computational representation of a given concept",
p. 445), pointing out that some flawed images produced by the computer
may be used to favor mathematical understanding and raise discussions.
Again, this perspective is interesting, but tends to suggest that such
'mistakes' or 'uncontrollable images' occur only in computer images, and
not in images generated with paper and pencil.

Examples like the one just presented regarding the tangent line frequently
appear in computer or graphing calculator environments; other examples
include different scales on each axis that make circles look like ellipses;
graphs that don't appear on the computer screen because the region of the
plane displayed is too small; graphs of functions with unexpected vertical
lines, etc. The unexpected answers are effects of the introduction of these
new actors in education. For some educators, those unexpected answers are a
disadvantage, or a source of erroneous or uncontrollable images; for us, they
represent an opportunity to discuss new mathematical questions, to favor
mathematical understanding, to produce interesting conjectures, and to
legitimize students' thinking that might be disregarded as wrong in other
media. New media, then, can provide a path for including more people in the
math classroom.

8.   VISUALIZATION, MEDIA AND THE VOICE OF THE STUDENTS

In Chapter 5, we described and discussed different views related to
visualization and media in mathematics and mathematics education, but we
did not present the voices of students giving their opinions about the role of
visualization and media in their learning. For this reason, we would now like
to show and discuss the opinions of some of the students who participated in
the teaching experiments conducted by Villarreal (1999), that we referred to
in the sections about derivatives and tangent lines. They used expressions
stressing visual aspects and expressed their thoughts and opinions regarding
media associated with mathematical learning. To make sense of the students'
assertions, the reader should keep in mind that they had little experience
with technology and used a DOS version of *Derive* during the teaching
experiments, which took place in 1997. We would like to present a synthesis
of the opinions of Mayra, Carolina, Camila and Maristela. The first three
students were introduced in previous sections.

> Mayra: When you learn just *writing, writing, writing*, I think it's harder
> that way for you to understand what the thing is. When you learn

graphically, there on the computer, it seems like . . . you know, it sheds a little more light on it. (p. 139)

Carolina: The graph on the computer is easier for you to get closer to what you want, except that then you have to *prove it in writing* ... you work in writing to see, you know, the tests. (p. 140)

Carolina: The advantage is this one of being able to *visualize*, and then get in, and the disadvantage, I think, too, is that often you aren't using your reasoning, but rather leaving it to the computer. (p. 141)

Maristela: ... the thing of just looking, without having to do calculations, I think helped me to think about the graph, understand? I think it helped me to reason more before going on to the calculations ... It's easier to get the concept like that, sketching the graphs. Now, a disadvantage is that when you need to do calculations, you have to stop, get the *paper*, because doing calculations all the time on the computer is kind of complicated . . . it's easier to use paper to do the calculations, you know?. ... I think that it's [a characteristic] of the computer, too, ... it's kind of boring to keep entering numbers in the computer, doing calculations, building the expressions in the computer. I think it takes longer than doing it by hand . . . there are a lot of symbols on the computer, and you can do it faster on *paper*. (p. 228)

Camila: ... but did you see that I wrote some things on *paper*? It's that I think, in order to think to create an expression before putting it on the computer, I think it's also good. . . to use *paper and pencil*. The computer is good for simplifying what you already know, and how you . . . like that time, it wasn't all very solid in my mind, so I thought it was important to write it on paper first, to afterwards transfer it [to the computer]. (p. 302)

The excerpts from Mayra, Carolina and Maristela suggest the awareness that students may have of the visual aspects of understanding mathematical concepts. These students' opinions are compatible with Zimmermann and Cunningham's (1991) definition of visualization in mathematics, wherein the authors refer to the use of images to obtain better mathematical understanding. One can also note the role the students assigned to paper and pencil. They use the paper-and-pencil medium to think-with before going to the computer (Camila); to later prove what they have seen on the computer screen (Carolina); or even to do calculations more rapidly than the computer, given the difficulties related to the software syntax when entering mathematical expressions (Maristela), in this case a DOS version of *Derive*.

Opposite opinions were offered regarding the use of computers and mathematical reasoning: for Carolina, reasoning may be lost with the computer, whereas for Maristela, the graphs on the computer helped her to reason. Camila expressed a deep-rooted view of the use of computers when she asserted that "the computer is good for simplifying what you already know"; in other words, we must know the contents before going to the computer. This position is common among teachers, and permeates many of the educational proposals using information and communication technology nowadays.

It could be concluded that the students have a traditional perspective regarding the use of computers in mathematics, that paper and pencil continue to be 'the medium' for studying and understanding mathematics, and that they consider computer-generated figures to be just useful aids for comprehension. Their opinions suggest how traditional media permeate their mathematical activity, in agreement with Lévy's (1993) assertion that a new medium doesn't supplant an old one, and support the notion of knowledge as being produced by collectives of humans-with-media.

Although many of the students expressed doubts about using the computers, we nonetheless observed the same students using the visual and algebraic resources of the software *Derive* intensely when they were engaged in mathematical tasks, suggesting that, although they may not be aware of it, the computers were playing an important role.

The attitudes and beliefs of the students may be influenced by their traditional mathematical instruction, which at the same time is permeated by the classical media: paper and pencil (or chalk and blackboard) and mathematical textbooks. Let us turn our attention towards mathematical textbooks in the next section.

9.   VISUALIZATION, EXPERIMENTATION AND BOOKS

In the preface to a classical Brazilian analysis textbook (Lima, 1976), the author made the following recommendation:

> A mathematics book should not be read as if it were a novel. You should have paper and pencil in hand to re-write, in your own words, each definition, the statement of each theorem, verify the details which are sometimes omitted in the examples and the demonstrations, and solve the exercises given with each topic studied. It is also convenient to draw figures (mainly graphs and functions) in order to attribute intuitive meaning to the reasoning in the text. Although figures do not intervene

directly in the logical argumentation, they serve as a guide to our imagination, suggest ideas, and help to understand the concepts. (p. VIII, our translation)

We want to emphasize two ideas in the above quote: 1) one learns mathematics with paper and pencil, and 2) visualization has an auxiliary role in mathematics, not being part of logical arguments. Humans-with-paper-and-pencil is the collective most commonly accepted in the mathematics community, with visualization relegated to the back seat. These ideas are certainly not new within the mathematics community, as discussed previously in Chapter 5; what is new is that, in this case, they are being stated explicitly to the students in the form of advice.

In spite of the fact that mathematicians do not consider figures to be part of logical arguments, we would like to show their importance to the construction and the understanding of proofs. In order to do so, we selected a classical theorem in calculus: the Mean Value Theorem. Following is a proof found in an old edition (fifth) of Thomas and Finney's *Calculus and Analytic Geometry* (1979, p. 155-157). In this proof, a figure plays an important role to obtain a better mathematical understanding of the theorem and its demonstration. The Mean Value Theorem states:

Let $y = f(x)$ be continuous on an interval $[a,b]$ with $a < b$, and differentiable on the interval $a < x < b$. Then there exists c such that $a < c < b$ and
$$f(b) - f(a) = f'(c)(b - a)$$

Previous to the proof, a graph and a brief explanation about it are shown. We reproduce them here as they appear on p.156-157 of Thomas and Finney (1979):

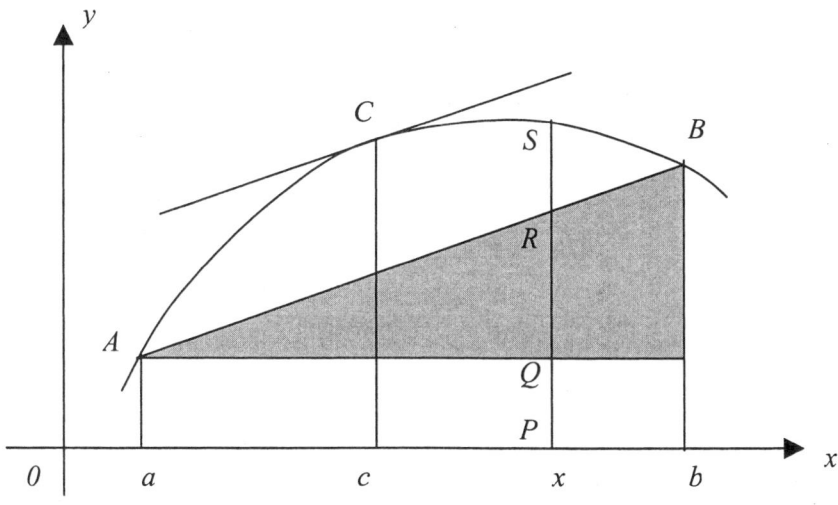

3-31 If $y = f(x)$ is differentiable on (a,b) and continuous on $[a,b]$, then the length RS is a function of x that is differentiable on (a,b) and continuous on $[a,b]$. In fact, RS satisfies all the requirements of Rolle's Theorem because it is zero at $x = a$ and $x = b$. Therefore, its derivative is zero at some value c of x between a and b. When we translate this observation into a statement about the function f, we obtain one of the most useful theorems of calculus.

Proof of the Mean Value Theorem. The vertical distance between the chord and the curve is measured by RS in Fig. 3-31 and

$$RS = PS - PR$$

Now PS is simply the ordinate y on the curve $y = f(x)$, so that

$$PS = f(x)$$

On the other hand, PR is the ordinate on the chord AB and may be found by using the equation

$$y - f(a) = m(x - a),$$

which is the equation of the straight line AB through the point $A[a, f(a)]$ with slope

$$m = \frac{f(b) - f(a)}{b - a}$$

That is, the ordinate y of the point R on the line AB is

$$y = PR = f(a) + \frac{f(b) - f(a)}{b - a}(x - a)$$

Hence

$$RS = f(x) - f(a) - \frac{f(b) - f(a)}{b - a}(x - a) \qquad (1)$$

measures the vertical displacement from the chord AB to the curve $y = f(x)$ for any x between a and b.

It will simplify the discussion somewhat to change the notation on the left side of Eq. (1) and replace RS by $F(x)$. That is,

$$F(x) = f(x) - f(a) - \frac{f(b) - f(a)}{b - a}(x - a) \qquad (2)$$

Then

$$F(a) = f(a) - f(a) - \frac{f(b) - f(a)}{b - a}(a - a) = 0$$

and

$$F(b) = f(b) - f(a) - \frac{f(b) - f(a)}{b - a}(b - a) = 0$$

so that this function $F(x)$ is zero at both $x = a$ and $x = b$. But since $f(x)$ and $x - a$ in Eq. (2) are continuous for $a \leq x \leq b$ and differentiable for $a < x < b$, and the other expressions on the right side of the equation are constants, the function $F(x)$ satisfies all the hypothesis of the Rolle's Theorem. Therefore its derivative must be zero at some place between a and b; that is,

$$F'(c) = 0 \qquad \text{for some } c, \qquad a < c < b \qquad (3a)$$

If we take the derivative of both sides of (2), we get

$$F'(x) = f'(x) - \frac{f(b) - f(a)}{b - a} \cdot \frac{d(x - a)}{dx}$$

and the result (3a) is equivalent to stating

$$f'(c) = \frac{f(b) - f(a)}{b - a} \qquad (3b)$$

or

$$f(b) - f(a) = f'(c)(b - a) \qquad (4)$$

which is what we wished to prove.

We note that (3b) states that the slope $f'(c)$ of the curve at $C[c, f(c)]$ is the same as the slope $[f(b) - f(a)] / (b - a)$ of the chord joining the point $A[a, f(a)]$ and $B[b, f(b)]$; this is a form that is easily recalled.

Note that the understanding of the proof depends on the figure presented initially. The authors explain the construction of the function $y = F(x)$, then it is shown that $y = F(x)$ verifies the hypothesis of Rolle's Theorem, whose application finally leads to the desired result. Thomas and Finney's book also presents an example where the value of c, whose existence is ensured by the theorem, is calculated. Furthermore, the authors analyze an example of a continuous function where the theorem is not valid because it does not verify the differentiability hypothesis. The Mean Value Theorem is central to the deduction of other important results in calculus, and that is why we believe its understanding, including graphical interpretation, is very important for students and for teachers, as well.

Now, let us have a look at the proof we found in an analysis textbook (Lang, 1969, p. 60), where the Mean Value Theorem is numbered as Theorem 1.

Proof. Let

$$g(x) = f(x) - \frac{f(b) - f(a)}{b - a}(x - a)$$

Then $g(b) = g(a) = f(a)$. We apply Lemma1 to g, and obtain Theorem 1.[32]

The graph shown in Figure 7-26, without any explanation, follows the proof:

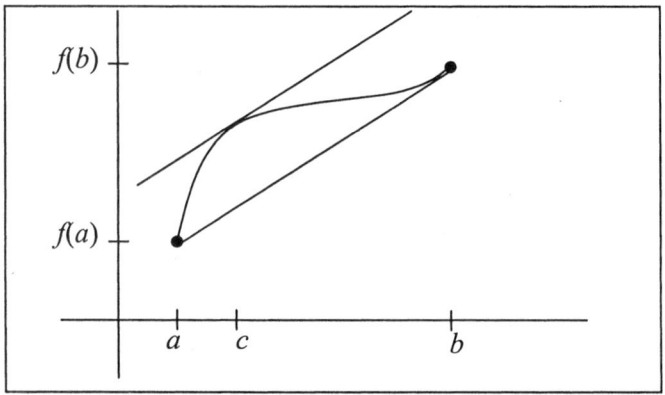

Figure 7-26. A graph in Lang (1969) that follows the proof of the Mean Value Theorem

The argument, in both proofs, is the construction of a function that verifies the hypothesis of Rolle's Theorem. Note that the functions are slightly different in each case, but in the first proof, we can **see** what the function $y = F(x)$ represents, and why that function verifies the hypothesis of Rolle's Theorem, and therefore makes it possible to get the desired result.

---

[32] Lemma 1: Let $[a,b]$ be an interval with $a < b$. Let f be continuous on $[a,b]$ and differentiable on the open interval $a < x < b$. Assume $f(a) = f(b)$. Then there exists c such that $a < c < b$ and $f'(c) = 0$.

The proof of the Mean Value Theorem in Lang's book is included as part of a quick review of calculus. The author warns that, although the book is self-contained, "it presupposes the mathematical maturity acquired by students who will ordinarily have had two years of calculus" (p. v), which is why the proof is included as a quick note, just to refresh the memory of those who studied the theorem at some point in the past. The proof has been compressed, and the student should know how to reverse that action. Thurston (1990) offered a nice description concerning these aspects:

> Mathematics is amazingly compressible: you may struggle a long time, step by step, to work through some process or idea. But once you really understand it and have the mental perspective to see it as a whole, there is often a tremendous mental compression. You can file it away, recall it quickly and completely when you need it, and use it as just one step in some other mental process. ... After mastering mathematical concepts, even after great effort, it becomes very hard to put oneself back in the frame of mind of someone to whom they are mysterious. (p. 847)

We feel the above quote could be applied to the written presentation of mathematical results in some books. Some books present 'zipped' versions of theorems, compressed in such a way that students may have to go to great effort to unzip them. Often the key to getting the unzipped version of some results is visualization.

Our aim is not to compare the books, which would not make sense considering that the first is a calculus book and the second, an analysis book, but to show the role that visualization may play in a mathematics textbook, even when the de-compressing process is left to the students. Lima (1976) appears to be suggesting that visualization is important and enlightening for students' understanding, but not official nor rigorous enough as a mathematical argument.

A counterpoint to Lima's statement, which could be considered extemporary, since his book was first published in 1976 at a time when there was little discussion about visualization in mathematics education, can be found in Larson, Hostetler and Edwards' *College Algebra* (1997), which shows a serious engagement with a particular graphical approach that we believe provides more democratic access to mathematics, as suggested by authors such as Kaput (1994) and Confrey and Smith (1994), among others. In the preface to their book, Larson, Hostetler and Edwards (1997) point out some features of the second edition. We would like to quote some of their comments referring to exploration and graphs.

Exploration Throughout the text, the Exploration features encourage active participation by students, strengthening their intuition and critical thinking skills by exploring mathematical concepts and discovering mathematical relationships. Using a variety of approaches -including visualization, verification, use of graphing utilities, pattern recognition, and modeling- students are encouraged to develop a conceptual understanding of theoretical topics. (p. X)

Graphics Visualization is a critical problem-solving skill. To encourage the development of this ability, the text has nearly 1500 figures in examples, exercises, and answers to exercises... (p. XI)

The above quotes reveal the value the authors attribute to visualization as an approach to develop mathematical understanding, or a skill to solve problems.

Books are also part of humans-with-media collectives, just as paper-and-pencil or computers are, and students also learn mathematics with them. Whether explicitly stated or not, textbooks are impregnated with the authors' conceptions regarding mathematics and the teaching and learning of mathematics. Lima values but attributes lower status to visualization, whereas Larson, Hostetler and Edwards (1997) consider it critical to problem-solving, and these differing perspectives are undoubtedly reflected in their respective books. The styles they choose to write and organize their books are conditioned by those conceptions.

It is difficult to imagine how a student could understand the Mean Value Theorem without a graph showing the construction of the function that led to its proof. We believe few students would be able to make sense of the demonstration without the figure. So, in response to those who share the perspective of Lima, the question arises: Must we forego understanding in order to have 'good mathematics'? We think not. Is it fair to create a dichotomy between visualization and correct logical development? We do not intend to respond to this question from the perspective of mathematics as a science, although it appears to us that traditional positions on the subject should be reconsidered. On the other hand, it seems clear to us that visualization and logical argumentation should be woven together in the classroom. If this already appears to be the case in traditional calculus textbooks, it will become even more so as information technology media become actors in thinking collectives that produce mathematics.

10. EXPERIMENTATION, VISUALIZATION AND REORGANIZATION OF THINKING

We have presented these examples to show aspects related to visualization, experimentation and media in mathematics education in different collectives of humans-with-media. We began by showing mathematical conjectures that were generated visually by collectives of university students from Brazil and Argentina using different information technology interfaces. The Brazilian students were biology majors, and the experimental-with-technology approach was used often as a pedagogical strategy in their regular mathematics course. The Argentinean students were pre-service mathematics teachers and participated in an extra-curricular experience where an experimental-with-technology approach was also the pedagogical strategy. High school students from Brazil also generated conjectures while working with unknown functions in a computational environment. Visual strategies were developed to validate the conjectures, and a discussion about the mathematical proof for different audiences was introduced.

The inclusion of interfaces, such as CBR, established new ways of exploring investigation, visualization and experimentation for eighth grade students from a Brazilian public school. The coordination of multiple representations, including body awareness, shows, among other things, alternative paths to introducing functions, avoiding an early passage through algebra. Moreover, there is some evidence, discussed in detail in Borba and Scheffer (in press), that a new kind of reorganization of thinking takes place, as sensors connected to body movements provide a very close connection between the awareness of these movements and standard mathematical representations like the Cartesian graph. Thinking about traditional physics graphs of distance × time is transformed after one experiences graphs generated through the CBR interfaces. Other paths, without using sensors, were also shown for introducing and developing topics such as functions and derivatives without first going through steps involving algebra and demonstrations.

The examples from teaching experiments conducted with biology majors showed: 1) a visual approach developed by a complex thinking collective to construct graphically the derivative of a function and 2) some aspects related to visual conflicts and the way the students overcame them. We discussed how information technology and writing, different kinds of technologies of intelligence, can generate 'uncontrollable images' for students. Media are neither good nor bad; they are part of collectives which produce knowledge, and they condition students at the same time that students go beyond the ideas of software designers, shaping the software to their own needs, in what was called an 'intershaping relationship' (see examples in Borba, 1993).

We listened to some students' voices expressing their opinions about the use of computers in mathematics, because they mentioned visualization as a main advantage associated with computers. They talked about paper-and-pencil and computers as means that can have some advantages, depending on the task. For us, their opinions reinforced Lévy's (1993) idea that the introduction of one medium does not suppress the existence of another. The students also expressed ideas about their own learning, and how different media (paper and pencil and computer with a DOS version of *Derive*) are more appropriate for a given task.

Finally, we decided to say a word about visualization and experimentation in mathematics textbooks, since they also form part of the humans-with-media collectives that are present in most educational settings.

If we return to the classical definitions of visualization reviewed in Chapter 5, and try to apply them to the examples presented in this chapter, it is our opinion that they fail to account for some aspects we pointed out, or address them only partially. The processes of visualization we have described were conditioned by the media; they were not performed by an individual, but by a collective of humans-with-media. If we look at the second part of Zazkis, Dubinsky and Dautermann's (1996) definition of visualization we read that: "...an act of visualization may consist of the construction, <u>on some external medium</u> such as paper, chalkboard or computer screen, of objects or events that the <u>individual</u> identifies with object(s) or process(es) <u>in</u> her or his mind" (p. 441, our emphasis). We emphasize the expression <u>on some external medium</u> because it implies that the medium is not an actor in the process of visualization, but a mere instrument to externalize objects or events associated with objects and processes in the mind. We disagree with the word <u>individual</u>, because we believe that, even when an individual is humanly alone, she/he is not 'really alone', in the sense that he/she is working together <u>with</u> some medium. Finally the word <u>in</u> remits us to the internal/external dichotomy. We can't say that the students' conjectures arose <u>in</u> their minds, or that the computer representations exist only <u>outside</u> their minds. The conjectures arose <u>with</u> the computer or graphing calculator representations. New mathematical problems arose <u>with</u> the technological limitations. Approaches, conjectures and representations were produced together with different technologies of information. The word 'with' deserves special attention. The main constructs of this book, humans-with-media and the reorganization of thinking, are related to the idea of 'thinking-with' that has been posed in Villarreal (1999) inspired by Lévy's notion of thinking collectives. We think with paper and pencil, with computers, with graphing calculators; we also think with a colleague. The fact is that we always think with media, producing a reorganization of the way we think, understand mathematics, make representations or solve problems. The idea that it is not only in collectives of humans, nor individually, that we come to know something, but with media, as well, is the key point of this book, and has been developed throughout the 90's by our research group (e.g. Borba, 1999b).

Visualization, experimentation, reorganization of thinking, and information technology are closely connected. Computer technology raised the possibility of 'laboratory-type' activities in mathematics education settings. Students were able to try coefficients of polynomial functions and construct several graphs in a very short time. Visual computer-based constructions in two or three dimensions are becoming increasingly popular, as well as generation of tables, and the study of patterns in them. Therefore, experimentation and visualization led to the discussion about multiple representations within mathematics education, and the possibility of introducing functions in middle school, high school and early college mathematics using not just algebra alone. This approach allows more students to deal with mathematics in a pluralistic way, democratizing access to mathematics understanding. Of course, multiple representations have generated new problems, such as difficulties in understanding specific results found in a given representation, and discrepancies between representations, although these can also lead to new conjectures that become new mathematics problems. Sometimes, as we have shown, even working with just one representation requires the coordination of different graphs, produced with the help of different media. In other words, intermedia coordination within a given representation is necessary. In our perspective, the experiences with computer technology, and the coordination of these experiences with other media, reorganizes thinking and transforms, in a recursive way, different humans-with-media collectives. In the next chapter, we will resume this discussion, considering information and *communication* technology, in particular the Internet, as a new actor in the production of knowledge.

Chapter 8

MATHEMATICS AND MATHEMATICS EDUCATION ON-LINE

1. HUMANS-WITH-INTERNET AND EDUCATION

In Chapter 2, we discussed the notion of reorganization of thinking, as different technologies of intelligence become a part of collectives that produce knowledge. Reorganization refers to what happens to human thinking, while the notion of humans-with-media emphasizes that this reorganization is so profound that we are unable to conceive of humans knowing without media. So far we have emphasized examples in which humans know with paper and pencil, or humans-with-computers know a piece of mathematics. We have stressed the role of sensors, graphing calculators, *Excel*, *Function Probe*, plotters and the like. We hope to have convinced the reader of the nature of the change that takes place when different media are associated with humans to produce knowledge.

In the same way we have seen how graphing calculators are transformed as interfaces like CBR are connected to them, we want to emphasize, in this chapter, the dramatic change that computers and education are experiencing with the connection of new interfaces like the Internet, and the www. The Internet can be thought as a physical interface, if one considers the cables that connect different computers and servers. If one thinks of 'bitnet' as an old interface for communication, one can understand how communication expands its possibilities with user-friendly software like the www. The Internet, understood from now on as all the physical and virtual devices that have allowed much of the world 'to be connected' since the mid-90's, i.e. less than ten years ago, has changed activities in all different kinds of fields.

From liberation politics in Chiapas, Mexico, to giant Nasdaq companies on Wall Street, the Internet has been changing the pace of our lives and our activities and the way profits are made, as analyzed by authors like Castells (1999, 2003). The list of changes could go on indefinitely, but in this chapter, we will focus on one area of change: the change that is taking place in (mathematics) education, in particular, in Brazil. The Internet is changing education in such a way that we are already often forced to characterize it using expressions such as 'face-to-face education' (in Portuguese, the literal translation would be presential education) and distance education.

Nonetheless, it seems that this change has not arrived with the necessary force in the mathematics education community. A look at the proceedings of the 2003 meeting of the International Group for the Psychology of Mathematics Education (Pateman, N.; Dougherty, B. and Zilliox, J. (Eds.), 2003) shows that very few papers dealt with distance education or effects of the Internet. Although there are exceptions, this suggests that the Internet has been used more by the mathematics education research community as a means to organize congresses, do bibliographical research, share information, or carry on discussions through e-mail lists, than as a subject for research. Does this mean that there is very little Internet in math education these days? We cannot answer this question yet, but in Brazil, due to its size and a system that fails to provide enough certified teachers for basic education, and the lack of openings in the universities for students, distance education (with heavy use of the Internet) has been a topic of heated debate. This debate has tended to place *face-to-face education* in opposition to *distance education*.

In the education community at large, positions have emerged that oppose the hastiness and superficiality of the distance courses compared to the face-to-face courses. The distance courses, they argue, are being developed only to hasten teacher education, and are a direct reflection of the policies of the World Bank. These were the arguments voiced in a debate throughout the state of São Paulo, during the years 2000 to 2002, regarding a one-and-a-half year course designed to grant college diplomas in pedagogy to elementary school teachers of the state public school system who had only high school diplomas. It was offered by important institutions in the field of education and research, but met with strong resistance from the faculty of the institutions involved, as well as others who were invited to participate in the project. The course was based on video-conferences featuring lectures that were transmitted to different groups in different locations. The teacher-students were allowed to propose activities, and it was possible for them to ask questions. In each tele-class, where the teacher-students were located, there were teaching assistants present to coordinate the activities. During holidays and on weekends, face-to-face activities took place. For proponents of the course, it was neither hasty nor superficial, and represented the only way for public school teachers who did not have a college diploma in pedagogy to earn one.

This brief description of a state-wide continuing education program for teachers serves to set the stage for raising the possibility that framing the debate in terms of face-to-face education vs. distance education may be inadequate. For example, the opportunity that these teacher-students had to hear lectures by, and interact with, professors who are well-known in the field of education, more-or-less directly via video-conference, is something that most students of pedagogy enrolled in face-to-face courses do not have. On the other hand, it can be said that one professor, giving a lecture to four classes of forty students each, will not be able to accommodate as many questions as would be possible in a face-to-face classroom of fifty students.

Nevertheless, it is an undeniable fact that the pedagogical debate persists. How did the professors and teaching assistants conduct the course? Did they use the established traditional educational approach, e.g. starting with theory and progressing to examples and exercises, in the case of mathematics, or did they employ alternative approaches that have characterized trends in mathematics education presented in this book[33]? The fact of being face-to-face or at a distance does not invalidate this discussion. Similar arguments can be cited that suggest that placing distance education in opposition to face-to-face education polarizes the debate inappropriately. A better question to ask is whether or not there are pedagogical approaches that are more appropriate for the www than others.

We subscribe to the notion that, just as a medium is neither good nor bad, face-to-face and distance education are neither good nor bad, either. We propose that research is needed to deconstruct this dichotomy, which currently defines the debate for many in Brazil. For this reason and others, members of GPIMEM began doing research in the area of distance mathematics education in 1999. Our first step was to become familiar with the possibilities offered (very few at the time in Brazil) and study the limited literature on the subject. For financial reasons, we discarded the possibility of using expensive environments for distance education developed abroad. We searched for free software and, only at the end of 2002, found a free environment, developed with public grants in Brazil, with the aim of promoting distance education: TelEduc[34]. Prior to that discovery, however, we started offering a distance course, in 2000, for mathematics teachers as

---

[33] For other trends in mathematics education, see the collection *Tendências em Educação Matemática*, published by Editora Autêntica. www.autenticaeditora.com.br

[34] For more information access http://teleduc.nied.unicamp.br

part of university extension work and also as a 'field for research' for the members of the group involved with this type of research[35]. During the early versions of the course, we used free chats and the infrastructure of UNESP to create an e-mail list and a homepage. The course, which is entitled *Trends in Mathematics Education*, has been offered four times since its inception, and includes discussions of Critical Math Education; Ethnomathematics; Modeling and Information Technology in Mathematics Education; Philosophy of Mathematics Education; teacher education; and in the last two versions of the course, issues related to fractals, Euclidean geometry, and functions.

In conjunction with studying technological possibilities of distance education, we were also looking for a means to ensure that a pedagogical perspective based on dialogical relationships would be embedded in the technological structure. We settled on a model in which the 'chat' ending up becoming the basic environment for the course. For three hours a week, during a period of one semester, twenty teachers, one technician, one teaching assistant (graduate students or undergraduate students of scientific initiation), and the professor of the course[36] participated in a virtual class in which pre-scheduled texts were discussed. Participants debated various aspects of questions raised by the professor as well as the students. In addition, participants took advantage of an e-mail list to exchange longer messages in-between sessions. At the end of the virtual part of the course, information about participants, summaries of the debates that took place in previous sessions, and questions posed for future courses were stored and made available on a site created for that purpose. Thus, the central aspect of the model used in the course was the guarantee of synchronous activities, such as the chat room where everyone 'met', and non-synchronous activities, in which each individual developed activities for the course on their own time, as in the case of the e-mail lists and bulletin boards. Since last year we have been able to use TelEduc, an environment that brings together the technical possibilities that were not previously available together[37].

Mathematics teachers from all over Brazil have participated in the extension courses over the last four years. A few of the participants were from other countries in Latin America. We have given priority to teachers who live far from São Paulo for two reasons. One was a research issue,

---

[35] Marcelo Borba, Geraldo Lima, Telma Gracias, Ana Paula Malheiros, Vânia Neves.

[36] Marcelo C. Borba

[37] For a detailed description of the model of the first course, see Gracias (2003) and Borba and Penteado (2001). For a broader debate of the various models used in distance education, see Valente (2003), Axt (2003), Maltempi (2003) and Moran (2003).

especially in the first course offered in 2000: we wanted to have the experience of interacting with teachers whose faces we did not know, as we believed this to be a component of this new modality of education. The second one was a socio-political one, as we wanted to contribute, albeit in a small way, to mitigate the concentration of knowledge in the State of São Paulo, which produces more than half of the research in a country that has a total of 27 states.

2. THE NATURE OF INTERACTION IN A DISTANCE EDUCATION COURSE

One of the first studies conducted by members of GPIMEM, led by Gracias, focussed on issues related to the nature of the interaction that takes place in courses like the one just described. Gracias (2003), based on theoretical ideas presented in this book, illustrates, with several examples, how reorganization of thinking takes place in distance courses conducted over the Internet. She points to aspects such as communication in network, and non-linearity and speed of communication, which have been previously discussed in other studies, but makes advances in her discussion, based on Lévy (1999), of how the chat room, which was the main vehicle of the course that she analyzed, serves as a space for creating meaning. She argues that the Internet contributes to a notion of space that is increasingly plastic, to the degree to which it introduces a new notion of proximity that is based on the interests of the participants in a virtual environment.

It is in this sense that the *multi-logue* described by Borba and Penteado (2001) and Gracias (2003) takes place. By *multi-logue*, we mean the occurrence of various intersecting dialogues, as takes place in chat rooms, where members are involved in various discussions simultaneously, and a given individual 'skips' from one discussion to another. It is this nature of the chat room that modifies the nature of the production of knowledge in this environment. It is different from the interaction in the classroom, considering that, for example, the professor of the distance course might be engaged in a discussion of modeling at the same time he is responding to another question of an administrative nature in a parallel dialogue.

Let's take an example from the *Trends in Mathematics Education* course described above. The topic of the day was the use of software and regular calculators in the mathematics classroom. Teachers were expected to have read some literature that had been sent to them previously. The dialogues appear below translated into English, although they were originally spoken-written in the chat in Portuguese, with the exception of the participation of one Spanish-speaking teacher from Argentina. In the transcripts, teachers are identified by capital letters, the teacher of the course as TEA, and the technician as TEC. The time is shown in brackets in the left margin. The dialogue may be difficult to follow at first reading, but the discussion that follows it will help to make things more clear. It may help to know that the bold face, italics, capital letters and the like were a methodological procedure we used in a first-level analysis as a means of identifying different 'conversational' themes that were taking place simultaneously. The original dialogue, in Portuguese, included typographical errors, abbreviations, and limits imposed by software on the number of words that could be written.

[20:36] <A>	*I am a little reluctant to use calculators and other technologies in case the students become too comfortable.*
[20:36] <B>	It depends on the projects we have. Lets return to the question raised in our last session. What are the problems that we want to confront?
[20:36] <C>	**I want what you promised our colleagues, too.**
[20:36] <D>	I UNDERSTAND YOU, F. MY MESSAGE IS THAT WE CAN USE THESE TECHNOLOGIES WITHOUT "PSYCHOLOGICAL PRESSURE". EVERYTHING IN ITS PLACE.
	[D is responding to an earlier statement by F: "Before, to get to the roots of a second degree equation, I wasted time . . . Today I enter the equation into a calculator, or a software, and it gives me the answer . .this adaptation scares me . . . we stop thinking, understand D?"]
[20:36] <TEA>	*and they think that media are agents, too, so I say that these media are associated with pedagogical approaches, because they are not innate beings, and that's why we discuss software design.*
[20:37] <TEC>	**That's OK, C...**
[20:37] <F>	*I agree with you, A.*
[20:38] <TEA>	*I didn't understand, A!*
[20:38] <G>	*Too comfortable, why?*

[20:38] <TEA>   *Media are agents too, and knowledge is not simply expressed by media; media are subjects, as well, and this is the thesis of my article, and various others.*

[20:38] <H>   *This is another point that we could return to now, Prof. Marcelo – discuss software design.*

[20:38] <A>   *My concern is how far we should go with media.*

[20:38] <C>   **M, how will this that you are suggesting help us in our teaching practice?**

[20:38] <I>   **Yeh, I want to, too.**

[20:38] <F >   *My concern is what media to use.*

[20:38] <TEA>   *I didn't understand, A!*

[20:38] <J>   B, in my situation, I don't think I even know anymore what problem we are confronting.

[20:38] <B>   *Related to this point, the association of information technology with pedagogical practice, it seemed to me that the evaluation of the experiences . . .*

[20:38] <TEC>   **Ok I...**

[20:38] <L>   *A, I agree with you to a certain point, but if you prepare a specific class for using the calculator, a really engaging class, the students will perceive its importance.*

[20:39] <N>   Means that even using media alone, I am learning.

[20:39] <B>   *done in working groups (H and B) were always very favorable.*

[20:40] <C>   *I didn't understand either, Marcelo [TEA].*

[20:40] <M>   **C, if you are referring to the demonstration of Fermat's Theorem, I think that it would modify what I accept as a valid demonstration in a mathematics course.**

[20:41] <A>   *OK, but in the next class without calculators, etc., they will ask for them, they will always want to use the technologies, and how can you use them every day?*

[20:41] \<C\>	HOW WOULD WE "STOP THINKING", IF IT IS THROUGH THINKING THAT WE COME TO KNOW TECHNOLOGY THOROUGHLY?
[20:41] \<G\>	WE DON'T HAVE TO STOP THINKING . . .
[20:41] \<N\>	WELL, ONE CAN EXECUTE SOMETHING, SEE IT, CONJECTURE, AND WITH THE HELP OF ANOTHER, VALIDATE IT.
[20:41] \<M\>	***What importance I place on symbolic manipulation, the structures, calculations, . . .***
[20:42] \<F \>	*Marcelo. In the readings, mathematical processes often appeared as research.*
[20:42] \<I\>	*And that's where the didactical contract comes in, the negotiation with the students. I always use calculators, and have few problems with it.*
[20:42] \<G\>	I AGREE.
[20:42] \<F\>	*We teachers don't have access to this type of software.*
[20:42] \<O\>	I DON'T THINK OF THE WORK AS BEING GUARDED BY AN INSTRUMENT OF CALCULATION.
[20:43] \<O\>	IN THE SENSE OF SAYING
[20:43] \<O\>	TODAY WITH THE INSTRUMENT, TOMORROW WITHOUT IT.
[20:43] \<L\>	*That's exactly what I wanted to say about this in my last sentence, A, that students perceive the importance of using calculators, but they also know to what point it helps and when it starts to "get in the way", although this might not be the right term.*
[20:43] \<TEA\>	*F, I didn't understand. Please explain.*
[20:43] \<G\>	THE USE OF THE CALCULATOR IS NOT RESTRICTED TO ITS MANIPULATION.
[20:43] \<O\>	TO AVOID GETTING INTO A BAD HABIT.
[20:44] \<O\>	I THINK THAT CONSTANT USE CAN CREATE SITUATIONS THAT FAVOR THE THOUGHTFUL USE OF TECHNOLOGY.

[20:44] <A>   *L, I don't really know to what point it helps or gets in the way.*

[20:44] <G>   WORSE IS THE FACT THAT OUR STUDENTS DON'T KNOW HOW TO MANIPULATE A SIMPLE CALCULATOR, OR EVEN INTERPRET THE RESULTS.

[20:44] <N>   IT DEPENDS ON WHAT THIS CONSTANT USE IS LIKE.

[20:44] <G>   I AGREE WITH O.

It can be said that there are five situations occurring at the same time in this piece of transcript. The first, in *italics*, refers to a discussion about the use of media in teaching practice. It involves the professor (TEA) and students A, F, G, H, I and L. Student B also becomes involved afterwards.

The part of the transcription in CAPITAL LETTERS refers to a discussion about the use of technology, whether or not it impedes 'thinking'. At first, students D, G, N, and O are involved, and later student C, as well.

The text that appears in arial font, (students B and J) and ***bold italics*** (students C and M) are dialogues between two students. They were apparently not very fruitful, as two students migrated to another discussion: B migrated to the discussion in *italics*, and C migrated to the discussion in CAPITAL LETTERS.

The situation in **bold** is a brief conversation between two students, C and I, and the technician (TEC). Student C soon migrates to the discussion in CAPITAL LETTERS, and student I migrates to the discussion in *italics*.

The data suggest that it is possible for debates on various themes to occur at the same time, and point to the speed with which new themes and questions arise. They also show how a student who is engaged in one dialogue or discussion one minute can move to another the next (the example of student C, who abandoned the dialogue with student M and moved to the discussion in CAPITAL LETTERS). There is also an example of an issue being raised that fails to generate debate, the question posed by student B: Lets return to the question raised in our last session. What are the problems that we want to confront? He posed a question that failed to raise a debate, and then moved to a more 'heated' discussion, the one in *italics*.

Although the example presented above may be long and difficult to follow, it helps to illustrate to the reader what it is like to experience this educational practice in this type of environment. Multiple interactions occur, and the professor has difficulty in following all the dialogues and deciding which one to participate in. The professor's interventions are also rushed, illustrating the pace of the class at some points, in contrast to other moments when three or four minutes go by with almost no intervention. Colors and 'faces' are used to indicate feelings that are sometimes communicated more easily in face-to-face conversations like in a classroom. But there is no temperature, there are no looks, and we don't know if there are gestures on the Internet within the design used for the course (without images in real time).

Although all the consequences of these differences are not yet clear, we also noted that the dialogue established follows rules that are different from those followed in a face-to-face class. Thus, various dialogues may be occurring at the same time, or as we proposed previously, a single debate re-organized around different rules, as opposed to a coordinated discussion. In a conventional class for graduate students, it is common for the professor to start the class with a lecture, followed by a period of questions and discussion, or to adapt a seminar model, in which the lecture serves as the basis for students' analysis and synthesis by the students together with the professor. In a distance course like the one described above, because of the way the chat room functions, a message sent by someone may generate multiple responses almost simultaneously, and lead other participants to have various interactions with the comments presented.

It is true that the same can occur in face-to-face classes, with the parallel conversations, looks, and gestures, but in a chat room environment, we can observe the interests of the sub-groups guiding various synchronous interactions. In the above transcription, one can also observe the fragmentation of time for each participant, who rather than following a linear debate, dealt with multiple experiences at the same time.

Thus, in this example, we observed how the notions of dialogue and time, in addition to the more obvious notion of space, are transformed when a collective of humans is joined together with information technology and its various interfaces. We understand that we are still in the pre-history of this new type of interaction made possible by the invention of new interfaces, and that, on the other hand, with the rapid progress in new technology, we will probably soon be in qualitatively different stages.

The analysis of the four courses offered to date also illustrates how the construct *humans-with-media*, proves to be appropriate as a starting point for understanding the production of knowledge in this environment, and also as a means of describing the type of change that has occurred in environments like the chat. This vision of the relationship between technology and knowledge is based on Lévy's (1993) analysis, in which the history of knowledge produced by humanity is permeated and conditioned by the different technologies of intelligence – orality, writing, and information technology – which we call media, to emphasize the communicational aspect. For Lévy, knowledge is never produced by humans alone, but always together with non-human actors. Technologies are created by humans, and are impregnated with humanity, and reciprocally, humans are impregnated with technology. The type of knowledge produced in societies where orality is the main instrument is distinct from that produced in societies that have some form of writing. Analogously, when we deal with a qualitatively different technology of intelligence, like computers, new collectives, composed of humans and non-humans, are formed. In other words, new collectives of humans-with-media constitute themselves as actors in this production of knowledge.

Identifying the role of new technologies in given thinking collectives has been the focus of a good part of the research carried out by GPIMEM in the ten years of its existence. Thus, we already discussed the way graphing calculators condition the production of knowledge in the mathematics classroom, as it would be unlikely to have occurred in the same way if the students had not used them (see, for example, Borba and Villareal, 1998). In these cases we would say that a collective of humans-with-graphing-calculator took form. Analogously, the constitution of this virtual space, formed by various courses that are offered to mathematics teachers, generates collectives that are only possible with the Internet, and the different interfaces used in the different courses. The Internet made it possible for students from all over Brazil, a country larger than the continental U.S.A., and from Argentina, to participate in virtual classes one night a week without having to travel to Rio Claro, São Paulo, one of the major centers of mathematics education in the country. Furthermore, the professor's choice of chat rooms, a resource made available thanks to the www interface, as the principle means of communication, allowed simultaneous dialogues to take place, which is not the case in, for example, video-conferences (which are also used in distance education) nor in face-to-face environments. In this sense, this particular humans-with-Internet-chat-rooms collective produces knowledge with its own dynamic.

It is not only the Internet that brings changes to the interaction; it is also the way the Internet is used by humans. The Internet definitely changes the realm of possibilities for teachers who do not live near centers that produce mathematics education, and it brings changes to the nature of interactions that take place.

3.   CHAT AND MATHEMATICS IN THE CLASSROOM

A typical class of this distance course for mathematics teachers (we will refer to them as students from now on) functioned as follows. Students would read pre-assigned texts beforehand; Borba and Penteado (2001), for example, to discuss the use of computer technology (software, graphing calculators, the Internet) in the classroom. The main concern of this particular text is the domestication of a new media, in which no changes take place in pedagogy, the role of the teacher, nor content, despite the introduction of new media actors. Two students would be chosen in the previous class to pose questions to get the discussion going. The teacher would make sure that at least some of the key issues of the assigned reading would be addressed.

This arrangement led to discussions like the one transcribed in the previous section. But what happens when mathematics is in center stage? In the four versions of the distance course, we have privileged the discussion about mathematics education. We believe this is very important, but we also believe it is important to do mathematics. As researchers in particular, we became increasingly interested in how mathematics would or would not be transformed as developed in the Internet environment. The third time the course was offered, in 2002, we did our first exploratory study on the subject, assigning one section to discuss Euclidean geometry and functions, using software for both. In the 2003 course, we developed two classes on these two topics and a third on fractals. Of course, we had no intention of teaching these topics to the teachers. The reasons were multiple: we believe that continuing education presupposes that teachers already know something about the topics, and also, we believe that 'teaching' should become an increasingly dialogical activity, as proposed by Freire and his followers. Secondly we have no intention of covering any topic in two three-hour sessions. Another reason is that we have no rigid syllabus, since we need to be flexible in order for teachers to explore mathematical issues at the level they prefer, since there were middle school, high school and university level teachers, and a few who taught at different levels at the same time, or had graduated recently and were looking for a job.

In the case of fractals, which was a new topic for most teachers, the arrangement for the class was more similar to the other classes, as a book had recently been published on fractals (Barbosa, 2002). The book included an explanation about fractals and suggestions regarding how to introduce it in the classroom, from elementary school to beginning university level.

When a problem from Euclidean geometry was posed, the reflections of one of the students called our attention. During the discussion, Eliane[38], said:

> I confess that, for the first time, I felt the need for a face-to-face meeting right away . . . it lacks eye-to-eye contact (class 6, April, 29th, 2003, 19:24)[39].

---

[38] Eliane Matesco Cristovão, High School teacher.

We still do not know how to interpret this sentence. But it appears to us that the discussion of a geometry problem intensified the need to share a blackboard and chalk, or scratch paper and pencil. But the debate did not continue, and she said nothing about blackboard or paper and pencil; we are only raising a weak conjecture. A somewhat stronger conjecture is that available technology does not support mathematical production, since we have no way to share drawings, or even to write algebra comfortably and conveniently. On the other hand, it is also possible that we are not open to dealing with new features in a new 'environment'. In other words, we may just not understand Internet in the way we believe we comprehend software like *Derive*. These are some of the issues that are still wide open for debate.

The class about functions in the 2003 course emphasized pre-assigned problems that were aimed to stimulate experimentation. We have given versions of the problems that were originally generated by the biology majors, as discussed in Chapter 7, and proposed another based on the technology we had available. We generated algebraically a graph like the one shown in Figure 8-1. The students, like the reader, could see the graph on the website of the course, but could not see the equation. The task was to guess the equation.

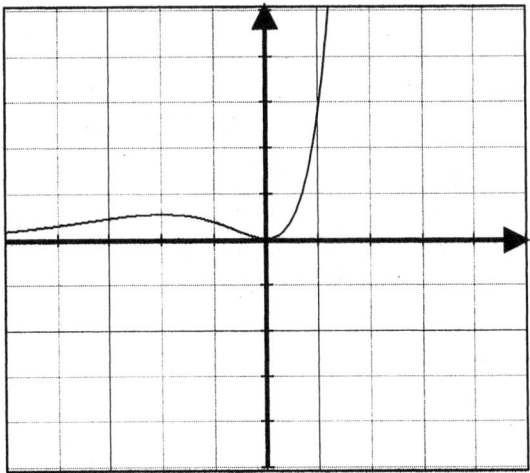

Figure 8-1. A graph to guess its algebraic expression.

The students discussed different problems and, similar to what happened in the multi-logue presented in the last section, different subgroups engaged in different discussions. However, at about 20:30, a focus on the function problem seemed to attract most of the students. The first guess was that the graph was a *"product of a quadratic times a trigonometric* [function]" (20:30). Another student suggests that *"we have one curve for x<0 and another for x>0"*, which we believe could be interpreted as referring to a piecewise-defined function. Most students were using *Winplot*[40], a free software for functions that has been translated into different languages, including Brazilian Portuguese. They were making their attempts, but there was no way to share what they had done, except sending an attachment file for the others to open, or having the technician upload a given solution to make it available on the homepage. Although these were reasonable bypasses, they were rarely used, and when they were, they could not maintain the pace of a real-time discussion.

After a while, a third student tried $y = x.sin(x).x^2$, and suggested that the first solution was incorrect. Nobody challenged the fact that there was an extra x in this equation compared to the first solution proposed: a product of a quadratic and a trigonometric function. But the discussion went in other directions; although the teacher could not offer hints that would give away the solution to the problem, he conceded that a quadratic function was part of the answer. After a few more hints, the teacher said that the answer was $y = x^2.e^x$. A nice discussion about the product of two well-known functions followed the intervention of the teacher. Students used quite a bit of intuitive language (as the teacher had previously) to show their understanding. For example, one said:

> ...it becomes greater than one [referring to the function]...making the product bigger and bigger than the value of the parabola itself, which does not happen before the y-axis, since the exponential is less than one, and although the product is increasing, it makes the parabola shorter (20:57).

We interpret this sentence as the student making sense of the behavior of the product of the two functions when x tends to $+\infty$ or $-\infty$. In the first case, since both functions are increasing and positive, and

$$\lim_{x \to +\infty} e^x = \lim_{x \to +\infty} x^2 = +\infty,$$

the product is also increasing and

---

[40] For more information about *Winplot* access http://math.exeter.edu/rparris/winplot.html

$$\lim_{x \to +\infty} x^2 . e^x = +\infty .$$

In the second case, a classical and common argument to explain the behavior of the product when $x \to -\infty$, is that the exponential function is going to zero faster than the quadratic function to $+\infty$ and that is why the product of the two functions tends to zero, since the exponential function is 'stronger' than the polynomial functions.

This mathematical discussion also led to considerations regarding how to make the arguments more formal in a classroom with teenagers or young adults, and how to keep track of students' steps when they investigate such a problem in the classroom. For the purpose of this book, we can say that what is important about this example, and about others that we are analyzing, is that we know little about mathematics produced by collectives that include the Internet. We could say that there is no difference between what happened in the example presented above and what would have happened in a regular class with *Winplot* and computers. In other words, there is no strong indication in our analysis that the Internet played as important a role as it did in the mathematics education discussion presented in the previous section.

In that example, and in many others we have collected, we have a plural discussion, with many subgroups engaging in different discussions according to their interests. We have not found examples like this in the part of courses that focus on mathematics[41] that would make clear the role of the Internet specifically for mathematics. Of course, we can say that there was a collective of humans-with-the Internet discussing mathematics, since if there were no Internet, those people would not have been discussing mathematics, being in different states of Brazil. But we cannot say (yet, and maybe never) that mathematics was transformed in the same way that the teaching of function has changed in classrooms with the introduction of graphing calculators and *Winplot*.

It can be said, however, that the participants of our distance courses were able to do mathematics and experience the experimental approach, although the sharing of the experience was somewhat more complicated. They were not able to show what they had on the graph. They could not point. They had to write without a 'proper' environment for writing mathematics, and maybe this is the difference, and this is the mathematics produced by collectives of humans-with-the Internet (in the environment we are using, without videoconferences). So we do have contingent answers, but in this instance,

---

[41] Although we are not comfortable with this distinction between mathematics and mathematics education, as we believe that they are in fact more intertwined than we make them appear when we make this distinction, we opted to leave it this way since we want to look for specificities of mathematics discussion in courses like the one under scrutiny.

unlike examples presented in other parts of the book, we do not have similar cases to confirm or discard these ideas, which leads us to state that more research is urgently needed. We consider this aspect of the research to be wide open, and our intention in reporting it, is to invite the reader to join us in this effort.

4.   RESEARCH PROBLEMS

In the next chapter, we will focus on how we have adopted different research procedures as different research questions emerge, and conversely, how we have needed to seek appropriate new questions as different procedures were used. Distance education, however, is a very peculiar case; first, because it is the last big challenge we have faced as a group, in terms of research methodology, understood in this book in a broader sense that includes a view of knowledge and the kind of question that is asked; second, because there are research issues that remain unsolved. Therefore, besides the need for more research to solve the question regarding the specific participation of the Internet in humans-with-internet-*Winplot* collectives that produce knowledge, we still have to discover how to go about it.

Doing research in distance education also raises new questions about qualitative research methodology in this virtual context. As shown in the first section, *multi-logues* call for different analysis and reporting procedures. The use of different fonts and different styles of letters becomes essential in order for most readers to follow the researchers' analysis, since the linear sequence of talk, as appears in the program that records the chat room conversations, doesn't allow one to make sense of what is being discussed most of the time.

If solutions to this problem, which is more of a procedural nature, were already found, it is not yet clear what this would mean in a 'natural environment', where the research is carried out. Qualitative research, and its predecessors from anthropology, emphasize the need to work in natural environments in education (Lincoln and Guba, 1985), but how should we then deal with the deconstruction of our experiences in time and space? What does 'natural' mean in the Internet? These are some of the questions that remain to be answered.

Up to this point, we have only used analyses of the transcriptions of chat room conversations and e-mail lists that, unlike common research recordings, actually do not need to be transcribed, since the talk is automatically transcribed.

But if we understand that the nature of the text produced is differentiated, that it is a mixture of speaking and writing, what consequences does the fact that this differentiation is not yet taken into consideration have for research? On the one hand, we use quotation marks when we speak of writing, or we use the expression *written-spoken* to indicate the fact that most of us think of chat as a space for writing, although we use all the informality of orality. But this just suggests that we are dealing with the notion of new oralities, in the same way that Lévy (1993) proposed that secondary orality is occurring when we read a text. In the case of the Internet, it may be secondary writing or tertiary orality, but we have not yet detailed the nature of this writing-orality.

Thus we have used the 'virtual electronic space' as our research environment, but what can be said about the twenty different physical spaces from where the participants access the site or the chat room of the course? Shouldn't we be investigating this, as well? Some have told us that they eat a sandwich during the class (Gracias, 2003), and others have proposed a toast, each with his/her glass of wine, in the final session of the course, as often happens at the end of a face-to-face course, and which would be more natural for those who are participating in the distance education course from PC's installed in their homes. What does an 'interview' by e-mail or chat room mean? How could triangulation be done, as proposed twenty years ago by Lincoln and Guba (1985), as a way of distancing our statements from mere opinion? These are open questions that are included here to provoke a collective debate on the themes. On the other hand, it could be that they indicate these researchers have fallen into the very trap they have tried to avoid, and want to reproduce with the Internet the same research practices they are accustomed to using without being open to new ones. This is yet another open question, that we hope will be taken up by some of the readers/teachers/researchers.

We are developing research in an attempt to solve the problems discussed in this chapter. We will also be investigating new technical solutions for some of the problems encountered. For instance, we are studying the possibilities of using applets that can be manipulated on the Internet that would allow everyone involved in a chat to see the results in real time. There is a 'math chat' being developed in Brazil that could solve part of the awkwardness of not being able do write an integral sign or raise an exponent in a chat. Of course, if we use these new interfaces, the humans-with-media collective will change, and humans will become different as well. So future researchers will have to deal with these issues.

Chapter 9

METHODOLOGY: AN INTERFACE BETWEEN EPISTEMOLOGY AND PROCEDURES

The theoretical perspectives, examples, and discussions that we have presented throughout this book show the systematization of the work developed by our research group (GPIMEM) since 1993. Due to the linear quality of the written medium that we chose to communicate our experience - a book - we elected the following order of presentation: first we addressed the central theoretical constructs, human-with-media and reorganization of thinking, and discussed modeling, experimental-with-technology approach and visualization; then we presented examples taken from our research projects, conducted individually or with other researchers from within as well as outside of GPIMEM. The examples were intended to shed light on the theoretical perspectives, and to raise new issues. However, this exercise in linearity obscures the paths that have led us to the particular landscape described in this book. In the act of research, there is no linearity, as there is in the act of communicating the results in writing.

Although the book was written by two GPIMEM members, there is a larger group of researchers that form a thinking collective that organizes itself around various projects that, although they may be individual (e.g. masters or doctoral research projects), they are always supported by the group in the study of texts or articles by new authors, the discussion of ideas, criticism, contributions or suggestions for research in progress, or the elaboration of new projects. Thus, the collective production of the group, and the non-linearity of the research, form part of the 'backstage' of the book, and for this reason, we propose that the reader travel in time to understand the context of the emergence of GPIMEM and its manner of conducting research in the Brazilian mathematics education scene.

GPIMEM began its activities in 1993, under the coordination of Marcelo Borba, and is composed of professors of the Graduate Program in Mathematics Education (UNESP – Rio Claro), a technician, undergraduate students in mathematics, and graduate students in mathematics education at UNESP. The students develop masters and doctoral research, as well as scientific initiation projects[42].Associated researchers from other institutions also develop investigation in collaboration with GPIMEM members. In addition to advising the students, the professors develop research associated with projects supported by funding agencies of the Brazilian government. The group emerged at a time when educational research in Brazil was characterized by the influence of so-called qualitative research (Lincoln and Guba, 1985) as a predominant tendency.

Brazilian educational research was, until the mid 80's, based on a positivist paradigm that valued measurement. As in many other parts of the world, investigators developed their research using experimental designs in which a measurable outcome, usually obtained through written tests, would be the result. About 15 to 20 years ago, as this trend began to lose popularity in Brazil (Fiorentini, 1995), the mathematics education movement gained momentum and was very influenced by the ideas of Freire (1976), the notion of ethnomathematics, whose birth is associated with names such as D'Ambrosio (2001), and a phenomenological approach to mathematics education (Bicudo, 1999). Freire and his notion of dialogue helped to establish the legitimacy of dialogue between researchers, teachers and students and hastened the deconstruction of the notion that objects of research are separate from the knower, in particular the researcher. Freire also always emphasized that it is not possible for a research design to be politically neutral and free of human influence - an assertion that implies the impossibility of having neutral questions and neutrality in science. In other words, what we choose to look at, and how we look at it, is unavoidably a result of our political point of view. As D'Ambrosio introduced the notion of ethnomathematics in Brazil, he invited the mathematics education community to consider and experience methods and views of knowledge that were drawn mainly from anthropology. Expressions such as 'field work', 'looking through the eyes of others' and 'comprehension' were imported from anthropology into the nascent Brazilian mathematics education community. That is why, not surprisingly, most of the research in this community is done with the strong influence of qualitative research (Denzin and Lincoln, 2000). Qualitative research stresses using open-ended

---

[42] Scientific initiation involves the development of research projects by undergraduate students under the guidance of professors-researchers. In practice, it functions as preparation for graduate school.

interviews, videotaping and other research procedures which allow the researcher to build a deeper understanding regarding the question chosen by the researcher, conditioned by the milieu where he or she is involved. In this epistemological perspective, knowledge is seen as contingent, as negotiated among different communities.

As a research group, GPIMEM has been investigating information and communication technology associated with mathematics education within the above framework. Since 1993, we have been researching the use of graphing calculators and computers with high school and undergraduate students. At first, the only resources we had for doing research in the classroom were graphing calculators, since neither schools nor the university had laboratories equipped with enough computers for an entire class of students. Borba started developing some classroom research in his mathematics course for biology majors, but initially, teaching experiments conducted with individual or pairs of students was the methodological option of various graduate students of the group at that time.

Teaching experiments (Cobb and Steffe, 1983; Confrey, 1991a; Borba, 1993; Borba and Confrey, 1996; Villarreal, 1999; Steffe and Thompson, 2000; Benedetti, 2003) can be seen as a sequence of meetings in which a researcher, interacting with students, takes a 'close look' at how students deal with tasks posed to them, or that they pose to themselves, and tries to model how students think about those tasks. For the most part, we used a variation of teaching experiments, as discussed by Cobb and Steffe (1983) and Steffe and Thompson (2000). We say that it is a variation of the teaching experiment because we share some of their concerns regarding the kind of observation that can be done in a setting where students are filmed and can be closely observed, although we did not develop long term experiments, for an entire semester or year, as Steffe and Thompson (2000) have suggested.

The initial work with teaching experiments in GPIMEM came as a transformation of the previous work developed by Borba (1993), in which he conducted teaching experiments with high school students to study the coordination of multiple representations of functions using *Function Probe*, a software designed for Macintosh only at the time. Visualization and experimentation were two key notions in his work. Associated with these was the notion of the intershaping relationship, in which users' ways of knowing are being shaped by computers at the same time users/knowers are shaping software in ways quite different from those intended by the designers. Visualization, experimentation, the intershaping relationship, and multiple representations are key concepts for GPIMEM, and later research deepened our understanding of them until they formed part of the base that gave rise to the notion of humans-with-media that anchors this book. Teaching experiments were very important in this process, as they allowed for a 'close look' at how students produced different knowledge with computer technology and with paper and pencil, stressing aspects related to the key concepts we have just mentioned.

This was the case of Souza (1996) who investigated the way in which high school students used graphing calculators to study quadratic functions, and presented a pedagogical proposal based on the experience. Villarreal (1999) conducted a detailed study of thinking processes of university students working with calculus concepts using the software *Derive*. In her study, she pointed out, among other things, the need to pay attention to the coordination of representations produced by different media, which she called *intermedia coordination* – a notion that was considered and examined in greater depth in later studies.

Simultaneously with the development of the fieldwork of the various research projects, the study of new authors (Tikhomirov, 1981; Lévy, 1993) and contacts with other researchers introduced new theories to the group, which permeate the research projects, the questions posed, and the methodology. Lévy's (1993) concept of a thinking collective, and his particular perspective which considers orality, the written word, and information technology as all being technologies of intelligence, together with Tikhomirov's (1981) discussion of the role of the computer as a mediator of human activity, were all central to the theoretical constructs presented in Chapter 2.

Among the various collaborations we have had with other researchers, we would like to highlight one in particular. Beginning in the mid 90's, we initiated a study involving new interfaces connected to computers or graphing calculators. In conjunction with TERC, a research institute in Boston, U.S.A, we started developing work with the LBM (Line Become Motion), a device created by the Math of Change Group at TERC[43]. This tool allows the movement of mini-cars along a track to be graphed in different formats (distance × time, velocity × time, acceleration × time, etc.) on a Mac computer; and conversely, a graph or an algebraic expression can 'drive' the mini-cars. Graphing calculators connected to sensors like CBRs (Calculator Based Ranger), which were also used in our research, can also generate a graph of someone's body movements. As interfaces such as LBM and CBRs became available, aspects which were not as salient before became more so. For instance, the discussion about the role of the body emerged as we observed how important, and at times difficult, it is for students to coordinate the movement of the body with the Cartesian graph. This question was addressed by Scheffer (2001), who conducted teaching

---

[43] For more information about TERC (Technical Education Research Center), access www.terc.edu

experiments with eighth-grade students, analyzing how they relate their body movements with the graphical representations as a function of time, produced by the CBR and LBM interfaces. Scheffer's research was informed by writings about conceptions regarding the body, and mathematical narratives that were brought to the group and discussed. It also led to a collaboration with Borba (Borba and Scheffer, 2003, in press) involving functions and multiple representations, presented previously in this book.

Following the same methodological line of the teaching experiment, and raising questions about the coordination of multiple representations of functions, Benedetti (2003) developed research with first-year high school students using a shareware software. Lévy's (1993) notion of the thinking collective is present in Benedetti's analyses, who worked with tasks that included functions unknown to the students and observed the way in which the software design conditioned the conduct of the tasks by the students, bringing new elements to the discussion of the intershaping relationship.

The teaching experiments developed by Borba (1993) and Souza (1996) were carried out with students working individually, whereas Villarreal (1999), Scheffer (2001) and Benedetti (2003) worked with pairs of students. It is important to note that even the transition from teaching experiments conducted with only one student to experiments carried out with pairs of students has to do with methodological choices that seek resonance with the epistemological position adopted by the group: that knowledge is constructed by a collective of humans-with-media, in which the interaction with another students is considered fundamental. This methodological option does not invalidate analyses of the individual work of the student, but rather extends the horizons of the research, exploiting some of the advantages of interaction between students, as suggested by Fontana and Frey (1994, 2000): richer data, greater stimulation for the participants, and mutual help between them. This makes it possible to obtain data more spontaneously, as the intervention of the researcher is reduced because the students dialogue naturally with their classmates, explaining what they are thinking.

A methodological option that is common to all the teaching experiments was the decision to videotape the meetings with the students. During the course of the different research projects, strategies for filming and analyzing the tapes were improved. What is the best way to film? How many cameras should be used? Is it advisable to have a technical assistant present, or could that be problematic? What steps should be followed in the analysis of the videotapes? Are complete transcripts necessary? If so, in what cases? Questions like these were discussed continually in the group. As a result, over the years, our research group has developed a sequence of steps that has been followed by most of the members when they analyze videotapes, which can be summarized as follows: watch the video between sessions, looking for key scenes that could result in 'emerging questions' in the next meetings; take notes after every session with the students, even if they have no apparent connection with the research question; after the teaching experiment is over, watch the videos again, looking for episodes that could shed light on the research question; transcribe these episodes; elaborate on them, looking for findings; check if notes or students' work provide counter evidence; present the episode and initial analysis to peers, members of GPIMEM; study possible alternative interpretations which have emerged; write the report.

The decision to conduct research with small groups of students, which could be called 'laboratory research', was also related to the limitations of the technological infrastructure available when GPIMEM was initiating its activities. The first studies conducted in the classroom were done using graphing calculators. These studies were developed in the afore mentioned mathematics course offered to undergraduate students in biology, taught by Borba since 1993 in the Biosciences Institute of UNESP, Rio Claro, Brazil. As already mentioned in Chapter 6, where the course was described in detail, two pedagogical strategies were employed in the course: experimental-with-calculator, and modeling. This transformed the course into a setting for research seeking answers to the question: How does the graphing calculator (and later, the computer, as well) alter the micro-culture of the classroom where an approach based on experimentation and modeling is being used? The classes were always videotaped whenever activities related to any of the strategies described above were being used. Various research questions related to modeling and experimentation emerged from these experiences, and the data were analyzed from various perspectives. Malheiros (2004) made a close examination of the modeling projects developed by the biology students over a period of ten years, analyzing the mathematics produced in the projects. Additional issues were discussed in this environment, leading to the publication of various articles: the use of the graphing calculator in the production of mathematical conjectures and realization of experiments; the search for biological explanations for mathematical results; and even the definition of criteria for evaluating modeling projects (Borba and Bovo, 2002; Borba, Meneghetti and Hermini, 1999, 1997). This research, developed around a mathematics course for biology students, shows how inquiry regarding modeling takes on new forms when information technologies are used.

Although the math class for biology majors has been the main setting for the classroom research of GPIMEM, other investigations about the use of technology in schools have also been conducted in different school settings to investigate various other issues. For instance, Zanin (1997) discusses how a software like *LOGO*[44] can be used in a school that makes computer resources available but is inflexible with respect to following the curriculum. Speculating, as there is no evidence in her study, Zanin (1997) attributes this rigidity, in part, to pressure from the parents. Provoked by this speculation, Da Silva (2000) developed the theme of the position of parents regarding the use of computers in their children's school. She interviewed mothers whose children were among the class of students studied by Zanin. In her study, no evidence was found to corroborate the suspicion raised by Zanin. However, Da Silva's findings pointed to the socialization of the parents by their children into the world of information technology, in a process that is the inverse of what is commonly found in education.

Araújo (2002) conducted part of her research in the classroom. She carried out an analysis of students' discussions in a calculus course for chemical engineering students in which mathematical modeling and information technology were employed in the professor's teaching approach. The software used in this case was *Maple*. Araújo conducted observations in the classroom and followed the modeling projects developed by two groups of students, composed of four students each, and video-taped most of the meetings held by the students outside the class. She concluded, with respect to the methodology, that following students' work in groups outside of the class can lead to perspectives other than those defended by Borba in his research on modeling, who used students' written and oral presentations as data. For example, her research shows that students' views of mathematics and 'real situations' can be very different from what we anticipate. In this particular case, for example, students opted to create an imaginary city instead of studying a real one.

Barbosa's (2001) research also focused on modeling, albeit associated with the education of future mathematics teachers. His interest centered on investigating how future mathematics teachers conceive of mathematical modeling, taking into consideration their previous experiences with it and their conceptions regarding mathematics and the teaching of mathematics. However, first he had to resolve a methodological issue: How to create a setting to develop the research when modeling is rarely addressed or employed as an approach in undergraduate mathematics programs in Brazil? The solution he found was to create a situation that provided him with the environment needed for the study. This situation, which Skovsmose and Borba (2004), in an article devoted to research methodology, call an 'arranged situation', was set up in an optional extension course entitled *Modeling and Mathematics Education* offered to undergraduate students of

---

[44] Programming software designed for education. For more information: www.nied.unicamp.br/sobre/links/logo.htm

mathematics at UNESP (Rio Claro). Three students enrolled in the course participated in the study, which linked modeling to teacher training.

In-service teachers were also the focus of an inquiry conducted by other GPIMEM members. Penteado Silva (1997) investigated teachers' relation with computers in the school, conducting extensive fieldwork in an elementary school (pre-school through fourth grade) at a particularly opportune moment, when computers were being introduced into the school. She discusses how the various actors in the school, e.g. administrators, teachers, and students, re-arranged themselves with the arrival of the 'computer actors'.

Research linked to the education and training of teachers gained momentum in 1997, when GPIMEM elaborated and launched the Information Technology in Education Project (*Projeto de Informática na Educação - PIE* [45]), which was born in response to a social demand coming from outside the university: the need to give support to the process of implementing information technology in the public schools in Brazil. Shortly before that, GPIMEM had made an important advance in terms of infrastructure and equipment, acquiring a laboratory equipped with twenty-two computers for carrying out its activities. The new computer laboratory made it possible to develop extension courses on Computers in Education for undergraduate and graduate students at UNESP, Rio Claro, and for public school teachers. The objective of the courses was to initiate a process of computer literacy, mainly for public school teachers who were planning to introduce computers in the classroom. Additional activities carried out as part of the PIE project included providing support for teachers' work in the schools, analyzing the performance and appropriateness of the educational software available on the market, elaborating an educational software, and offering access to our computer laboratory, with the technical support of monitors, so that school teachers in Rio Claro could participate in a distance course via the Internet.

The development of multiple activities by the PIE project also resulted in reflections on research methodology. One particular challenge for GPIMEM, for example, was the interest expressed by two public school teachers in using computers in their mathematics classes (fifth to eighth grade), when the school had only four computers available. On the one hand, there was the teachers' interest and the possibility of conducting research about the use of computers in the school; on the other hand, however, the equipment

---

[45] The Information Technology in Education Project was made possible by the financial support resulting from an agreement between UNESP and IBM-Brazil. More details on the project can be found in Penteado and Borba (2000).

available at the school was insufficient. How could one research a nonexistent situation? The solution at the time was to create an arranged situation (Skovsmose and Borba, 2004) that would approximate the imagined situation: the GPIMEM laboratory became the setting for the mathematics classes administered by the teachers. The students walked from their school to the laboratory at the university, where their teacher then taught their math class. In this way, the arranged situation provided data to inform an analysis of the imagined situation. Thus, from extension work that initially aimed to focus on computer literacy, new research questions emerged: How would teachers with almost no previous contact with computers deal with software in their mathematics classes? This research question brought new challenges from a methodological standpoint: How to carry out this action-research? By collaborating with the teachers? What would it mean to collaborate with the teachers? How can an environment be created to develop the research, since the school has no computers?

The setting generated by the PIE project, which was closely linked to the continuing education of teachers, inspired the later development of various studies and projects related to this subject. Da Silva (1999) studied conceptions of undergraduate mathematics students and math teachers of elementary and middle schools regarding the use of computers in the mathematics classroom. With a proposal that extended beyond mere research, Penteado initiated a project, called *Rede Interlink* (Interlink Network) that involves a network of teachers working collaboratively with researchers and teachers-in-training to organize and elaborate activities for the classroom using information technology resources. In this context, Cancian (2001) discusses changes in the thinking and practice of teachers engaged in collaborative work using information technology in mathematics education. More recently, also inspired by continuing education for teachers being carried out within the PIE project, but with a focus on Distance Education, Gracias (2003) studied the nature of the reorganization of thinking in a course entitled *Trends in Mathematics Education*, coordinated by Borba and conducted entirely over the Internet, as discussed in the preceding chapter. Research on distance education via the Internet in particular, and the influence of the Internet on mathematics education in general, has been developed by the group since 1999, and represents an increasingly popular line of research within the group. These new research domains have also brought new problems and methodologies, as presented in detail in Borba (2004), and discussed in the previous chapter.

The objective of presenting the above report on the procedures and themes of the research developed by GPIMEM from its inception was to illustrate the conception of integrated research that permeates the group. It is our understanding that, in order to comprehend a phenomenon, such as the presence of information technology and the role of media in (mathematics) education, it is necessary to compose a network of research actions, as we have done. In this network, various issues associated with the phenomenon are addressed:

- The characterization of processes of knowledge production by thinking collectives of humans-with-media, considered as epistemological subject.
- Modeling as a didactic approach that is in resonance with the epistemological construct of humans-with-media.
- Teacher education and its conceptions related to modeling and the introduction of information and communication technology into daily school activities.
- Distance education and the influence of the Internet on mathematics education.

These themes, which make up the knots in our research network, are interwoven with other knots of a broader research network developed by other groups and individuals, so that we can discuss and understand the diversity of results and their relation to the types of questions and research methodologies adopted.

In many of the studies developed by the group, the research questions that guided the inquiries underwent modifications, based on the subsequent immersion of the researcher in fieldwork, or the need to limit or reformulate the question due to its magnitude or the impossibility of addressing the initial question. The methodological options associated with the questions were also modified as needed, so that the research design was not predetermined, but was emergent in nature.

Thus, we believe that in the research carried out by GPIMEM, the questions and methodologies emerge in an integrated manner, without a chronological order. In this sense, a version of a research question inspires a given research methodology, which in turn influences the generation of a second version of the question. There is, however, a search for coherence; if our questions revolved around investigating whether or not information technology improves the teaching and learning of mathematics, and if we believed this could be measured using traditional tests, we would have to seek other research methodologies, typically quantitative, that would provide ways of measuring this alleged improvement, even if, eventually, qualitative methods might be used.

On the other hand, we believe that the studies carried out by GPIMEM are characterized by a coherence among them that is due to the view of knowledge they all share. This view values understanding, not correct answers, and affirms that knowledge is produced by collectives composed of human and non-human actors, which we have called humans-with-media.

When we say that we value understanding, we mean that we are concerned with the processes students follow in their mathematical activities, with the procedures they elaborate, and with the particularities of their strategies, whether they are mathematically correct or not. Thus, we are interested in hearing the 'student's voice', and trying to understand it, which can certainly lead to changes in our perspectives. In this sense, the work of Confrey (1994) has informed our view related to knowledge.

Here it is important to return to Lincoln and Guba's (1985) notion of resonance, which emphasizes the coherence between a view of knowledge, research procedures, and pedagogy, and which was introduced in Chapter 3, where we extended the concept in a discussion of modeling as a pedagogical approach that is in resonance with the use of information and communication technologies. Lincoln and Guba (1985) cite, as an example, how a behaviorist view of knowledge would be consistent with research procedures that emphasize the use of tests and statistical analyses, just as epistemological views that emphasize understanding would be in harmony with qualitative methods that give importance to the way the students think and not the results obtained. We extend this 'harmony' between the various elements of research to the very nature of the questions posed, and as suggested by Lincoln and Guba, to the very pedagogy that we choose to research.

Of course there is no one-to-one correspondence between epistemology, research questions, procedures, pedagogy and so on, but there are limits. For instance, an aseptic view of knowledge, in which it is seen as not influenced by humans or technology, based on measurable variables such as test outputs, would try to overlook, minimize, or control for most of the interplay previously discussed.

Different views of mathematics and mathematics education can also lead to different approaches which may or may not be consistent with the above discussion. A view of mathematics in which there is just one path to building mathematical knowledge, and only one way of teaching, would hardly be consistent with our vision of knowledge, as it fails to take into account the changing media, among other factors. If mathematics is seen as a rigid body of unquestionable truths, then there is no possibility of discussing the influence of media in the mathematics produced by students[46]. Mathematics education based on this view of mathematics, or on the view that all schools need to do is transmit mathematics that has already been developed, would also be incoherent.

---

[46] For a discussion about this point, see Kaput (1994).

Upon analyzing the data, the notion of humans-with-media also assumes, for many of us, an important role, to the degree to which we seek manifestations of media that we judge to be relevant for a certain thinking collective at a given moment. Thus, in a given study, we show how a graphing calculator becomes imperative for the development of a given conjecture by a humans-with-technology collective; or, alternatively, we argue that the use of the Internet was only marginally important, and that it did not participate effectively in a given episode, considering that it could easily have occurred in the same way without this medium.

It is in this way that we believe that a theoretical perspective, like the one outlined in this book, can become important for those who want to do research, as they must seek the resonance discussed to avoid producing incoherent research. It is also important for those who are inclined to investigate the day-to-day practice in schools, since these theoretical studies can serve as guides so that new information and communication technologies are not used only as instruments to improve local, regional, and national test results. The arrival of a qualitatively different medium, like computers, must contribute to the modification of established traditional teaching practices.

We chose to raise questions of a methodological nature near the end of the book for various reasons. First, so that the reader would have the chance to become familiar with our theoretical position regarding technology, as well as a sample of some of our results. Secondly, because we wanted to emphasize that linearity exists only in the form of presentation, and since in many publications, methodology appears at the beginning or near the middle, we resolved to put it at the end. For us, research questions, procedures, and one's view of knowledge emerge together, and not in a given order. Another reason is that we wanted, after presenting various examples in different educational settings, to show how the research process has an extremely strong collective aspect, and to do this, it was necessary for the reader to know more about the work of our group.

In this sense, we as a group are one example of a collective of humans-with-media. Different humans, close to 100, have passed through the group or are still members of it. Different information and communication technologies have joined different collectives in specific research projects. We have changed the design of media, suggested changes for new designs of machines, and in particular, we have developed pedagogical activities based on our research. We have learned intensively from each other. We are a specific case of the notion of 'inverse socialization' (Da Silva, 2000); we have all socialized each other into different pieces of software, different theories, etc, regardless of whether the one who was sharing his/her expertise was a graduate student or an experienced researcher. Moreover, this collective supports us individually to keep up with the growing changes and the increasing pace as we go through different experiences in collectives with different information and communication media. GPIMEM is a collective of humans-with-media that produces knowledge expressed in books, articles, conferences, e-mail lists, homepages and in all different kinds of notes - oral notes, written notes and computer notes.

Yet there is another reason why it is important to develop research collectively: tackling a problem like the possibilities of change in mathematics education as new media enter our lives may seem very specific for some, but it is too broad to be taken on by one person alone. Triangulation, an idea from Lincoln and Guba (1985), is considered to be one means of improving the reliability of qualitative research, as it mitigates the influence of one given person or one given procedure. Having triangulation over the long term requires a team. In the examples presented in this book, we can say that there is triangulation with respect to procedures, as classroom research, interviews and teaching experiments were used when we are tackling specific research questions. Exploratory studies, in which we were just looking for focus, or to understand a given interface, were also developed. We have triangulation with respect to the research settings, as we developed research in various environments. We have triangulation regarding the media we research, as we do not want to be associated with just one software, or one interface. And finally, we also have triangulation in terms of the researchers involved, as close to one hundred people cooperated at different times, or at the same time but in different physical spaces (the Internet), to analyze data, offer alternative interpretations, and generate the theoretical constructs, view of technology, and pedagogical perspectives described in this book.

During these ten years, research questions, readings, theories, new ideas, procedures and results shaped each other and shaped us, and also shaped the way we think about information and communication technology. We believe that a broader view of methodology ranges from the specific procedures and steps taken in developing research to the epistemological discussion of the nature of the knowledge we have produced.

Chapter 10

POLITICAL DIMENSIONS OF INFORMATION AND COMMUNICATION TECHNOLOGY

Throughout this book, we have tried to highlight the changes that information and communication technology have brought to collectives of humans-with-media in action. Historically, the various technologies of intelligence have caused substantial changes of a varying nature, in the workplace, industry, politics, commerce, and education. If one travels back through time, it can be seen how education was generally reserved for the privileged classes of society, for the elite, throughout the history of humanity. For example, the cultural opportunities available to the urban population surpassed, and still surpass, those of rural populations, especially in poorer countries. Roche (1996) refers to this contrast in 18th Century France, where in the face of the predominantly oral, visual, and gestural culture of the rural areas, the city appeared as a milieu impregnated with writing, where many things appeared in print, and people obtained information from these printed materials: "The city constructs an original cultural universe, where writing represents some role, even for those who are unable to decipher it" (p. 177). In this historical context, the author indicates that the practices of writing "affect and shape the consciousness of a wide public. In the city, the common man, more and more, must read" (p. 199). Reading can be viewed as secondary orality, according to Lévy (1993). This secondary orality, which derives from reading what has been written, implies changes in the customs and habits of city dwellers.

Roche does not specifically analyze writing practices in the school, and for this reason, we would now like to return our attention to this subject, in particular, to the media used in these school practices, and the changes that they produce in the classroom. We approach the subject initially considering the introduction of the notebook into the classroom, and how it influenced school activities. In France, for example, use of the notebook became common in high school (10 to 14 year-olds) in the 16th Century, and was obligatory in the teaching of calligraphy in the 17th Century, but its generalized use in the elementary schools dates from the first third of the 19th Century (Hébrard, 2001). The cost of paper was a determining factor that limited its use until the more advanced grades, restricting children's learning in the area of literacy to reading. Hébrard reports that, around 1833, the use of the notebook in elementary education was considered by the Ministry of Elementary Instruction to be a sign of pedagogical modernity. Certainly these historical data differ from place to place. In Argentina, for example, the notebook was introduced into the classroom around 1920, nearly a century after its introduction in France, and was closely linked to the so-called new school movement, which recommended the use of a single class notebook as an organizing tool for school work (Gvirtz, 1999).

Both Hébrard (2001) and Gvirtz (1999) analyze writing in the school, and the particular physical support provided by the notebook, in different eras and societies. Both authors point to the work of British Anthropologist Jack Goody as a basis for asserting, like Lévy (1993), that writing re-structures our thinking. Gvirtz (1999) suggests that the notions of accumulation and productivity link the organization of school work with the principles of industrial work in effect at the beginning of the 20th Century. Up until then, the writing practices were carried out with small, portable chalkboards. Photographs of classrooms from that period show the walls covered with chalkboards and the students busy doing calculations, dictation, etc. When notebooks were introduced into the schools, everyday life inside as well as outside the school was transformed. These transformations affected the activities of the students. According to Hébrard (2001), with the use of the notebook:

> The student discovers not only how to organize the bi-dimensional space proper to the graphical order, but also how, through the writing, to control the time of his tasks and his days. Mixing text, schemes, pictures and even images with text, he gives himself the means of disposing of a proper instrument for organizing the encyclopedia of his knowledge. (p. 115)

The author analyzes in detail the writing practices found in notebooks in 19th Century France, which go beyond the simple fact of learning how to write. The use of the notebook implied not only knowing how to copy, write dictation, do arithmetic exercises, or solve arithmetic problems, but also know how to organize and present them, making the notebook a "small theater of school knowledge" (Hébrard, 2001, p. 137). The notebook thus constituted a daily register and chronology of all school activity, turning it into a notebook-appointment calendar, and was also transformed into a link between school and the family: parents could follow the progress of their children. It was also a medium for the school inspectors to supervise the teacher's performance. The French author's analysis allows us to suggest that the presence of the notebook in the school introduced a series of changes into the day-to-day activities in the classroom, whose importance equals that of broader administrative and curricular changes. This aspect was raised by Gvirtz (1999), who asserts that:

> The notebook is not a mere physical support ... On the contrary, it is a device whose articulation generates effects: in more concrete terms, the notebook constitutes, together with other elements, a shaper of the classroom (p. 160).

Gvirtz points out that, around the 1880's, the first discussions were reported regarding the use of paper in the schools in Argentina. These discussions emerged at a time when paper was no longer a luxury because of its lower cost and increased availability. As occurred in France, economic issues were linked to pedagogical decisions. For these reasons, learning to write with paper was reserved for the upper school levels. This brought on the debate between reformist pedagogues of the new school movement, who proposed the use of paper in the schools, and the defenders of the chalkboard. This debate continued until the 1940's, when the focus changed to other issues. According to Gvirtz, considering the changes introduced by the notebook, its use cannot be seen as a simple change in the technology used to report school activities, but as a re-organizer of life in the classroom.

We have included this example of the introduction of the notebook into the schools because many of the issues seem to parallel those raised in this book regarding resistance to using computers. Notebooks were not available to everyone. This is also the case when political dimensions of the use of computers are considered. Humans-with-notebook were foreign to schools at that time, similar to the way that humans-with-computers are not present in many educational settings even in the XXI Century. Similarly, a political divide existed in those days between humans-with-notebook collectives and the collectives of humans-without-notebooks. In the same way, it will still take a long time until humans-with-computers are a possibility in most schools or other educational settings; and the longer it takes, the faster the political divide will increase. Notebooks and the new style of school were concentrated in cities, creating a division that remains to this day in most countries. Computers are concentrated in some areas, as well, although the division is not exactly the same.

Castells (1999) is one author who argues that, more important than our traditional notion of space, is the new virtual one, in which the flux of information defines who belongs to a social group, and that the Internet is the very basis for this new notion of space. This argument adds support to Borba's (2002) assertion that the main justification for having computers in the schools is related to rights of citizenship. *Access* to information and communication technology is the key word. In France, two centuries ago, access to notebooks was restricted. In Brazil, not more than 15% of the entire population of almost 180 million people has access to computers, and if we consider access to the Internet, this percentage decreases. Even less available is fast Internet access, and access to a decent amount of software and support to use them. From our perspective, the main reason to invest in computer technology in schools is because they are public centers, which is a key issue particularly in countries where most of the population cannot afford to have computers at home. In countries like Brazil, where the minimum wage is about EUR 70 a month, and a new computer costs EUR 700 (with just basic software), it is important to have public computers. Even for those who can afford to buy a computer, they are not likely to be able to afford technical support, a fast Internet connection, updated software, etc. in a country where a 'well paid' (compared to the rest of the population) professor makes little more than a thousand euros a month, and a high school teacher earns, on the average, four hundred euros a month.

In a more recent book, Castells (2003) presents studies showing that the Internet was basically developed through market forces. This neo-liberal way of development implies that the Internet tended to 'follow the money', being available first in areas with higher concentrations of wealth, and helping to bring more money to those areas. In the U.S.A, for example, providers and Internet access are more highly concentrated in California and New York states. Within New York, there is a bigger concentration in Manhattan, and within this island there is a greater concentration of Internet access in the Wall Street area. Similar results are shown by Castells (2003) for other countries, such as Brazil, which has the same 'Internet geography' as the United States, more highly concentrated in wealthier areas. We believe that public policies are needed to take the necessary infrastructure to places that are less interesting to the market.

Expanding access, and improving the quality of access, is fundamental if we are to make real the 'dream' of Lévy (1993, 1999) of expanding the possibilities of democracy with more direct control and decision-making power using direct voting that would bypass the need for representatives. The Internet, which is the physical structure for this and, at the same time, the interface that transformed the potential of computers, has become the key word for this discussion. Without good access, you are out of the network of communication, and even your access to free software is denied. Free (or cheap) software are key to increasing access to the net. In Brazil today, the number of machines purchased by the public sector could have been doubled if they had not opted for the 'basic package' of *Windows* and *Office*. Some public entities that opted for Linux have experienced some difficulties in terms of access to specific software; for example, availability of mathematics software in Portuguese that runs in Linux is very limited.

If access to information and communication technology is key to full citizenship in order to avoid the creation of legions of socially-excluded people, as many authors claim, it should also be recognized that access alone is not enough. If access to fast Internet and good quality software were provided, would this guarantee inclusion and prevent new divisions between the haves and the have-nots? Hardly, would be our answer. Computers per se bring change to education, but this does not exclude the pedagogical discussion. How to match its various uses with other media in different educational settings? We believe that access to computers without pedagogical discussion can result in their incidental use, or in their domestication (Borba, 1999b), in the sense that they would be used in the same ways other media were used before. If we use computers as if they were a fast version of paper and pencil, we would be domesticating the possibilities of this new medium. We should propose pedagogical approaches that have synergy with information and communication technology. Notebooks shaped knowledge in the school in different ways than did the portable blackboards each student carried and had to erase in order to write more. Computers can provoke even greater changes, particularly if views of pedagogy and epistemology like those defended in this book are developed and gain favor within the community of mathematics educators, educators and policy makers. It is in this sense that the issue of information and communication technology in the schools is a political issue. Views of knowledge, and pedagogies that support change in school mathematics at various levels, can help us face, in a small way, the challenges of full democratization of information and communication technology.

The research we develop at GPIMEM gains this political dimension if one considers the fact that most of our research is developed in free public universities and public schools. We are not neutral in this regard, as we want to comprehend the insertion of computers in collectives of humans-with-media. A political concern, which motivated some of our research when we began our activities ten years ago, was to fight the dominance of the absolutist discourse in mathematics and mathematics education (e.g. Skovsmose, 1994; Alrø and Skovsmose, 2002). This has been one of the objectives of the ethnomathematics movement (D'Ambrosio, 1985; Borba, 1987, 1997a), which developed the idea that mathematics is culturally bound, and argues that mathematics developed within academia is culturally-bound, as well. The notion that knowledge is a product of humans-with-media supports the idea that knowledge is not only bounded by subject(s), but is also shaped by different media that/who are actors in the generation of socially-accepted mathematics, for instance. With theoretical postures like these, which are based in research and have lead to new investigations, another political dimension of research becomes apparent. We conduct research with a political agenda – a non-partisan agenda, but our research about modeling, computers in mathematics education, and the notions of visualization and experimentation in this book are also embedded in the political agenda of giving voice to students and teachers. We believe that the notion of just one mathematics, that everyone is expected to learn in the same way, silences the different voices of students and teachers who are involved in education. It was not by accident that our focus in the book has been on what students (and teachers) can do, instead of what they cannot do, their misconceptions, or their faults. Our agenda of making the classroom plural, emphasizing visualization and experimentation, is similar to Mellin-Olsen's (1987), who claims that students lack of access to mathematics was a political act. He was struggling against examinations and pedagogy that excluded students then, and still do today. Besides carrying on his cause, we want to present possibilities that interaction with new media can offer. New collectives of humans-with-media can develop different mathematics, and they should be accepted at all levels of mathematics education.

In a similar sense, we believe that some of the work we have been developing in GPIMEM is an example of the type of public policy that should be developed. Distance education courses, for example, like the one we have been developing for five years at GPIMEM, with strong emphasis on dialogue and interaction, becomes a pedagogical approach for this type of education. We should use the Internet to bring research that is developed at UNESP in Rio Claro closer to other areas that want to interact to us. Conversely, we should use the Internet to learn from other centers. We believe, therefore, that being part of the collective intelligence is a political act, and it is part of what public institutions like UNESP should be doing.

In spite of our political agenda, our intention is not to make our research into a manifesto. This is why we have also discussed problems we have encountered in implementing modeling as a pedagogical perspective, and we hope to have avoided any propaganda in favor of computers. This is the reason we have tried to triangulate most of our findings, either among members of the group, or within the literature of the area. In the same way we have tried to be critical about our own findings, we have also tried to be critical of the literature. As we do not believe we have the final word on the topics discussed in this book, we hope it provokes more debate, so we can proceed with our contingent certainties and working hypotheses about mathematics education.

To make this small contribution to mathematics education, developing research for ten years in environments disconnected from the 'center' of the production of knowledge located in the North America and Europe, we hardly had an alternative other than forming a research group like GPIMEM within our graduate program. Difficulties we confronted in implementing this kind of research in Brazil ranged from doing research about technology with just one Macintosh computer, two graphing calculators and two old PC's in 1993, to the challenges discussed in Chapter 8 regarding research methodology and the insertion of mathematics in distance education in Internet-based courses. Developing extension courses for parents from poor neighborhoods and teachers from the elementary, middle and high schools in Brazil, who earn little money, without all the infrastructure for research, is both a challenge and a political act again. This extension work became part of our research, and as we found funding for research, it was shared with students in need of financial assistance. It is a political act to face the challenge of providing continuing education for teachers in Brazil, even in the form of a modest (in size) distance education course. In such courses, we do research, aiming to implement large-scale programs, and provide access for teachers from all over the country to 'direct' contact with professors from one of the main centers of mathematics education in the country. Moreover, in this type of course, it is possible to invite guests (on line) from different centers. In the courses we offered, Ubiratan D'Ambrosio from Brazil, and Arthur Powell from the U.S.A., have been among the guests who have interacted with the teachers and graduate students enrolled in the course.

Maintaining a research group that focuses on information and communication technology in Brazil has been a political act in itself, in that it represents a rejection of expectations to conduct 'Third World research'. Maintaining it for more than a decade has been an act of citizenship, or an act of political resistance, in a world order wherein the division of labor between countries of the 'North' and the 'South' has been imposed by various means. In this 'world order', countries like Brazil are expected to conduct research only on hunger and deforestation, and not on information and communication technology, and to export oranges, not airplanes. The existence of our group is a 'grain of sand' to change this kind of logic. Nonetheless, networking does occur between the North and the South as well, which helps to confront this international division of labor in a small way. Our collaboration with Ricardo Nemirovsky's group at TERC, Boston, U.S.A., and with Ole Skovsmose's colleagues at the Centre for Research in Learning Mathematics, Denmark, provides examples of mutual profit from the collaborative research we develop. Collaboration between south-south, such as the one that resulted in this book, is also a way of strengthening mathematics education in both countries.

In any case, some of the difficulties of developing research within GPIMEM, and with its associated researchers at different levels, may be shared by other groups around the world. For instance, we have noticed how tensions emerge within the group as different 'times' provoke discrepancies. GPIMEM has a history of collaboration among its core members that can transform into an example of the collective intelligence proposed by Lévy (1998). We take advantage of the different qualities of our members, and the technology available, to compose an intelligence that designs and develops research - an intelligence that is greater than the sum of its individual parts, as discussed in Chapter 9.

On the other hand, we face tensions that arise between our different personal rhythms and the increasingly frenetic pace of technology development. The time that members of the group have to incorporate new interfaces and software varies. However, none of us has a psychological sense of time that keeps up with the 'technology market time', even considering that technology for (mathematics) education seems to lag far behind technology in areas such as communication and the military. This creates a tension in research: how to develop cutting edge research on technology, if before we have even grown accustomed to a given software, a new interface appears that transforms the possibilities of research and teaching and learning? How to coordinate research on 'old' information technology with research on 'new' information and communication technology? We do not yet know enough about how students learn with dynamic geometry software such as *Cabri*, *Geometricks* or *Sketchpad*, which have been the subjects of intense, worldwide research for more than a decade, and we have to face challenges of connecting sensors to calculators, possibilities of the www for mathematics education, applets, hyperdocuments, videopapers and so on.

We have dealt with the dilemma between developing more superficial research with the latest technology available, or developing profound research with technology that is scarcely used anymore, by coordinating actions among members of the group. We have members developing research with the usual kind of software in dynamic geometry, and we have other members exploring cutting edge developments available to us, even if we do not know if it is worthwhile developing research on it, to say nothing of the appropriate research question to be asked. We have also tried to 'filter' issues and findings from one study and see if they hold true in others. The construct of humans-with-media has helped in this regard, as it sets the stage for documenting the specific role of new interfaces, software and devices in a collective that constructs knowledge. Experimentation and modeling, and the associated notions of visualization and multiple representations, are notions that have helped us to cope with the tension between our internal sense of time and the pace of technology development, as they seem to be consonant with the development of computer technology, as we have argued throughout this book. Modeling, experimentation, visualization and multiple representations let students (and us) explore different topics at different paces, mitigating the tension between our time and technology-development time. Maybe even time can be seen as having political implications, as it frames social forces and the possibilities of research.

There are other kinds of time that also provoke tensions among group members and in the whole dynamic of the group. For instance, how to coordinate the tensions discussed above with the deadlines and demands of academia? Universities worldwide have been increasingly inspired by notions of time and productivity that come from industry, requiring a certain amount of publications, just as industries set production goals. There is also the time of social transformation and the need for rapid change in education. How to coordinate this rapid pace with the slower pace of qualitative research, which seeks to avoid being fooled by false certainties based on tests that unnaturally limit the scope of, and predetermine, the subject under investigation. We have no answers to these questions, but we invite the reader to maintain this dialogue with us, in various forums, showing the political underpinnings of your research, the research questions, the findings, the research methodology, and your doubts and uncertainties.

Afterword

The notion of *humans-with-media*, central in Marcelo Borba's and Mónica Villarreal's study, brings together two ideas: that cognition is not an individual but a social undertaking; and that cognition includes tools. These ideas draw on many resources in different fields; for instance, with reference to Niels Bohr and theoretical physics, it has been suggested that what we can grasp about nature depends on how we can approach nature, i.e. on the tools we have available.

In the following I shall address humans-with-media from an epistemological, an educational, and a socio-political perspective. From an *epistemological perspective*, humans-with-media points towards more fundamental conditions for producing human knowledge and understanding. One important point is that 'humans' is in the plural. This need not be so, and in fact several classic epistemologies think of the epistemic subject (i.e. the subject who might come to know) as a 'lonely' one. One example is René Descartes' doubting subject, who, in the middle of his or her doubt, grasped that *cogito, ergo sum* cannot be doubted. From this axiom, then, Descartes proceeded to develop all forms of knowledge. This represents an extreme form of epistemic individualism, which has become part of Western philosophy ever since. Epistemic individualism is also expressed in classic empiricism, as formulated by John Locke, George Berkeley and David Hume. There is no point in claiming that one can share a sensorial experience with others (my headache is certainly *my* headache), and if all knowledge comes through sense experiences, then the epistemic subject must be an individual. This individualism is recapitulated by Jean Piaget's genetic epistemology, combining rationalism and empiricism; and it has undergone a captivating reformulation through Ernst von Glasersfeld's presentation of radical constructivism. One has to construct knowledge from one's own experiences. Common to all branches of epistemic individualism is that the learner, the 'carrier' of knowledge, is an individual. This is supposed to be a basic condition for human understanding. It is a universal epistemic claim. So, when Borba and Villarreal talk about 'humans' in plural they challenge a main trend in epistemology and in the conception of learning. They suggest an epistemic collectivism. Learning takes place through processes of interaction.

This collectivism can be related to a Vygotskian framework, but also to a more general one, not assuming too much of the dialectal materialism, which Vygotsky did embrace. Epistemic collectivism brings about the claim that any interpretation of 'learning', 'coming to know' and 'knowledge', must refer to processes of interaction. Learning cannot be thought of as an individual undertaking but as a social one. In mathematics, epistemic collectivism is nicely illustrated by the format of Imre Lakatos' book, *Proofs and Refutations*, which makes an account of the dynamics of mathematical knowledge production through a fictitious dialogue. The teacher does not present pre-established knowledge in some condensed form, nor does any contribution from individual students present an all-embracing insight. Instead, it is the network of contributions, which is the 'subject' of knowledge production.

'Media' within the expression humans-with-media is also plural. A long tradition in epistemology has recognized the relevance of media for coming-to-know, although the notion normally used is 'tools' and not 'media'. (I shall switch rather freely between the two concepts.) Coming-to-know is not an operation carried out with 'empty hands'. It is, as Bohr also observed, a process including tools. This insight comes naturally, when one thinks of practices as a resource for knowledge. One is doing something: harvesting, cooking, experimenting, etc., and in such processes, we most often use tools. Tools could be considered auxiliary to the process of coming-to-know, as technological equipment can be thought of as supplementary for humankind. However, tools can also be thought of as essential to a practice, as defining the practice, and therefore as defining both content and forms of knowledge. This is pointed out when humans-with-media is considered a unit for considerations of knowledge and learning.

In many cases certain tools are taken for granted. For instance, is our knowledge of geometry based on tools? Some tools, such as paper and pencil, we might think of as so obvious that we cannot conceive of geometric knowledge separated from them. Generally speaking: an understanding of the developing of human knowledge always has to be related to the tools that might be at hand. Tools are not extrinsic to the development of knowledge, but an intrinsic part of the way we conceptualize. A fuller understanding of knowledge development must address the nature of tools which are at hand at the given time. At present, this establishes the computer as an important epistemological category.

Let me now look at humans-with-media from an *educational perspective*. This means that we do not try to outline basic categories for learning, but to reveal possibilities for educational initiatives and priorities. It becomes important to consider organizations of learning where students get the opportunity to work in groups and to negotiate. It becomes important to consider problem-based and projects-organized learning. How is the teacher-student communication organized and facilitated within a certain educational setting? In general: as 'humans' are plural, the educational perspective of humans-with-media makes us search for potentials with respect to communication and interaction.

The educational interpretation of humans-with-media also brings our attention to the media involved. Computers are introduced in the classroom practice, and what could that mean? How could communication, interaction and, eventually, the learning of mathematics be facilitated through this media? As an educational concept, humans-with-media serves as an invitation to reconsider the potentials of different tools that might be included in mathematics education. And, certainly, we do not have to do with simply establishing new motivational strategies. The question is instead: Do some tools have an essential impact on the qualities of learning in mathematics? Tools cannot be thought of as external facilitators of a process of learning, which, would proceed towards pre-established aims anyway. Tools are intrinsic to the nature of coming-to-know, and Borba and Villarreal provide a clear insight into the qualities which computers might bring to the learning of mathematics.

Finally, let me consider humans-with-media from a *socio-political perspective*. Tools could be distributed very differently depending on the context of learning, also when we have to do with mathematics education. How is access to computers, this powerful tool, distributed around the world? Who has access to it? According to statistics (see, for instance, UNESCO. *Education for All: Statistical Assessment 2000*. Paris: UNESCO, http://unesdoc.unesco.org/images/0012/001204/120472e.pdf) the population of school children from the so-called developed countries add up to 10% of the world's population of children, while 16% of the world's population of children do not go to school. It seems all too well documented that getting access to computers is a privilege for only a minority of the world's learners (being primary, secondary, tertiary, university or adult students). The prevailing discourse addressing mathematics education and computers, however, does not address this issue. Instead, it takes classrooms with computers for granted. This discursively constructed classroom comes to operate as a proto-type within mathematics education research. But when we consider the majority of sites for learning in mathematics, the reality is far from any prototype.

The notion of humans-with-media underlines the importance of addressing the questions: If computers can be seen as a powerful tool for learning mathematics, what does this observation then mean for the majority of learners of mathematics who have no access to computers? Are they barred from learning with certain qualities? Such questions bring forward the problem of exclusion and inclusion, which is touched upon in the final chapter of Borba's and Villarreal's study. It has been pointed out several times that mathematics education might serve very different socio-political functions, one being that it serves as gate keeper, providing access to possibilities and social advancement for some, while excluding others. Mathematics education could mean either empowerment or disempowerment. Does a computerization of mathematics education establish new forms of exclusion and inclusion? This raises the issue of citizenship and equality through education.

Let me point out a few more specific aspects of the socio-political dimension of humans-with-media. First, one could try to engage in a politics of providing access to computers for everybody. However, such an approach also raises questions concerning economic interests and business priorities. Companies provide offers for education, including software packages. But, as pointed out, media can mean many things, also when we consider computers. There can be many different kinds of software and different ways of handling these. Being a powerful tool does not mean that any educational interpretation and application of this media can be recommended.

Second, even though there are no computers in sight for the majority of the students of this world, it still makes sense to consider what theoretical insight could be extracted from studying computer-affluent environments, with the particular aim of applying this insight to contexts without computers. Could there be some insight about experimentation and visualization that can be useful also in computer-deprived learning environments?

Third, we must remember that any tool includes a restructuring of both learning context and the content of what is learned. This means that mathematics learned without computers need not be the same as the mathematics learned when computers are in place. Does the computer have some impact on the quality of learning, not only when we try to identify qualities in mathematical terms, but also in terms of reflection and critique? Reflections could, for instance, address the reliability of bringing mathematical tools into operation with respect to certain problems and tasks. Could computers help students to challenge an exaggerated trust in numbers or an ideology of certainty? Such questions become important when we address to what extent a computer environment in mathematics education might support the development of a critical citizenship.

Ole Skovsmose
August 2004

References

Alrø, H., and Skovsmose, O., 2002, *Dialogue and Learning in Mathematics Education*, Kluwer Academic Publishers, Dordrecht.

Anastácio, M., 1990, *Considerações sobre a Modelagem Matemática e a Educação Matemática*. Master thesis. Universidade Estadual Paulista. Rio Claro. Brazil.

Araújo, J., 2002, *Cálculo, Tecnologias e Modelagem Matemática: as Discussões dos Alunos*. Doctoral dissertation. Universidade Estadual Paulista. Rio Claro. Brazil.

Aspinwall, L., Shaw, K., and Presmeg, N., 1997, Uncontrollable mental imagery: graphical connections between a function and its derivative, *Educational Studies in Mathematics*. 33 (3) , 301-317.

Axt, M., 2003, Educação (a Distância): Apontamentos para pensar modos de habitar a sala de aula, *Revista Interface, Comunica, Saúde, Educação*. 7: 143-145.

Baldino, R., 1999, Pesquisa-ação para formação de professores: leitura sintomal de relatórios, in: *Pesquisa em Educação Matemática: Concepções e Perspectivas*, M.A.V. Bicudo, ed., Editora UNESP, São Paulo, pp. 221-245.

Barbosa, J., 2001, *Modelagem Matemática: Concepções e Experiências de Futuros Professores*. Doctoral dissertation. Universidade Estadual Paulista. Rio Claro. Brazil.

Barbosa, R., 2002, *Descobrindo a geometria fractal*, Autêntica Editora, Belo Horizonte.

Barton, B., 2002, O desenvolvimento de um registro matemático maori, *Boletim de Educação Matemática – BOLEMA*. **17**: 71-82.

Barwise, J., and Etchemendy, J., 1991, Visual information and valid reasoning, in: *Visualization in teaching and learning Mathematics*, W. Zimmermann, and S. Cunningham S., eds., Mathematical Association of America, Washington, DC, pp. 9-24.

Bassanezi, R., 1994, Modelling as a teaching-learning strategy, *For the Learning of Mathematics*. **14** (2): 31-35.

Bassanezi, R., 2002, *Ensino-Aprendizagem com Modelagem Matemática: uma Nova Estratégia*, Editora Contexto, São Paulo.

Ben-Chaim, D., Lappan, G., and Houang, R., 1989, The role of visualization in the middle school mathematics curriculum, *Focus on learning problems in Mathematics*. **11**(1): 49-60.

Benedetti, F., 2003, *Funções, Software Gráfico e Coletivos Pensantes*. Master thesis. Universidade Estadual Paulista. Rio Claro. Brazil.

218

Bicudo, I., 2002, Demonstração em Matemática, *Boletim de Educação Matemática-BOLEMA*. **18**: 79-90.

Bicudo, M., 1979, Intersubjetividade e Educação, *Revista Didática*. **15**. São Paulo.

Bicudo, M., 1999, Filosofia da Educação Matemática: um enfoque fenomenológico, in: *Pesquisa em Educação Matemática: Concepções e Perspectivas*, M.A.V. Bicudo, ed., Editora UNESP, São Paulo, pp. 21-43.

Bishop, A., 1989, Review of research on visualization in mathematics education, *Focus on learning problems in Mathematics*. **11**(1): 7-16.

Blum, W., 1991, Applications and modelling in mathematics teaching - A review of arguments and instructional aspects, in: *Teaching of mathematical modelling and applications*, M. Niss, W. Blum, and I. Huntley, eds., Ellis Horwood, pp. 10-29.

BOLEMA, 2002, *Boletim de Educação Matemática*, 18, M. Borba, ed., UNESP, Rio Claro.

Borba, M., 1987/1994, *Um Estudo de Etnomatemática: sua Incorporação na Elaboração de uma Proposta Pedagógica para o "Núcleo-Escola" da Favela da Vila Nogueira-São Quirino*. Master dissertation. Universidade Estadual Paulista. Rio Claro. Brazil. Published by Associação de Professores de Matemática (APM). Portugal, 1994.

Borba, M., 1990, Ethnomathematics and education, *For the Learning of Mathematics*. **12**(1): 39-43.

Borba, M., 1993a, Etnomatemática e a cultura da sala de aula. *Educação Matemática em Revista*. **I** (1): 43-58.

Borba, M., 1993/1994, *Students' Understanding of Ttransformations of Functions using Multi-representational Software*. Doctoral dissertation. Cornell University. Published by Associação de Professores de Matemática (APM). Portugal, 1994.

Borba, M., 1994, A model for students' understanding in a multi-representational environment, in: *Proceedings of the PME 18*, J. P. Ponte, and J. F. Matos, eds., **2**: 104-111.

Borba, M., 1995a, Overcoming limits of software tools: a student's solution for a problem involving transformation of functions, in: *Proceedings of the PME 19*, L. Meira, and D. Carraher, eds., **2**: 248-255.

Borba, M., 1995b, Funções, representações múltiplas e visualização na Educação Matemática, in: *Anais do I Seminário Internacional de Educação Matemática*. Rio de Janeiro. IM-UFRJ, pp.71-90.

Borba, M., 1997a, Ethnomathematics and education, in: *Ethnomathematics. Challenging eurocentrism in mathematics education*, A. Powell, and M. Frankenstein, eds., SUNY Press, Albany, pp. 261-272.

Borba, M., 1997b, Graphing calculator, functions and reorganization of the classroom, in: *Proceedings of Working Group 16 at ICME 8: the role of technology in the Mathematics classroom*, M. Borba, T. Souza, B. Hudson, J. Fey, eds., Gráfica Cruzeiro, Rio Claro, pp. 53-62.

Borba, M., 1999a, *Calculadoras gráficas e Educação Matemática*. E. Kaufman, and F. Cohen, eds., Série reflexão em Educação Matemática, **6**, Universidade Santa Úrsula, Rio de Janeiro.

Borba, M., 1999b, Tecnologias informáticas na Educação Matemática e reorganização do pensamento, in: *Pesquisa em Educação Matemática: concepções e perspectivas*, M. A.V. Bicudo, ed., Editora UNESP, São Paulo, pp. 285-295.

Borba, M., 2002, O computador é a solução: mas qual é o problema?, in: *Formação docente: rupturas e possibilidades*, A. Severino, and I. Fazenda, eds., Papirus Editora, Campinas, pp. 141-161.

Borba, M., 2004, Dimensões da educação matemática a distância, in: *Educação Matemática: Pesquisa em Movimento*, M.A.V. Bicudo, and M. Borba, eds., Cortez Editora, São Paulo, pp. 296-317.

Borba, M., and Bovo, A., 2002, A modelagem em sala de aula de matemática: interdisciplinaridade e pesquisa em biologia, *Revista de Educação Matemática* - SBEM. **8**(6): 27-34.

Borba, M., and Confrey, J., 1996, A student's construction of transformations of functions in a multiple representational environment, *Educational Studies in Mathematics*. **31**(3): 319-337.

Borba, M., Meneghetti, R., and Hermini, H., 1997, Modelagem, calculadora gráfica e interdisciplinaridade na sala de aula de um curso de Ciências Biológicas, *Revista de Educação Matemática* - SBEM. **5**(3): 63-70.

Borba, M., Meneghetti, R., and Hermini, H., 1999, Estabelecendo critérios para avaliação do curso de modelagem em sala de aula: estudo de um caso em um curso de Ciências Biológicas, in: *Calculadoras gráficas e Educação Matemática*, E. Kaufman, and F. Cohen, eds., Universidade Santa Úrsula, Rio de Janeiro, pp. 95-113.

Borba, M., and Penteado, M., 2001, *Informática e Educação Matemática*, Editora Autêntica, Belo Horizonte.

Borba, M., and Scheffer, N., 2003, Sensors, body, technology and multiple representations, in: *Proceedings of the PME 27 and PME-NA 25*, N. Pateman, B. Dougherty, and J. Zilliox, eds, **1**: 121-126.

Borba, M., and Scheffer, N., in press, Coordination of multiple representations and body awareness. *Educational Studies in Mathematics*.

Borba, M., Souza, T., Hudson, B., and Fey, J. (Eds.), 1997, *The Role of Technology in the Mathematics Classroom*, 8th International Congress of Mathematical Education, Gráfica Cruzeiro, Rio Claro.

Borba, M., and Villarreal, M., 1998, Graphing calculators and reorganization of thinking: the transition from functions to derivative, in: *Proceedings of the 22 PME*, A. Oliver, and K. Newstead, eds., **2**: 135-143.

Borwein, J., Borwein, P., Girgensohn, R., and Parnes, S., 1995, Experimental mathematics: A discussion. *Proceedings of the Organic Mathematics Workshop*. Simon Fraser University (October 28, 2002); http://www.cecm.sfu.ca/projects/OMP/

Borwein, J., Borwein, P, Corless, R., Jörgenson, L., and Sinclair, N., 1995, What is organic mathematics? *Proceedings of the Organic Mathematics Workshop*. Simon Fraser University (October 28, 2002); http://www.cecm.sfu.ca/projects/OMP/

Brugger, W., 1969, *Dicionário de Filosofia*, Editora Herder, São Paulo, Brazilian version of the 6th ed. of *Philosophisches Wörterbuch*, published in 1957 by Verlag Herder and Co. Freiburg im Breisgau. Germany.

Buerk, D., 1990, Writing in mathematics, a vehicle for development and empowerment, in: *Using writing to teach mathematics*, A. Serrett, ed., MAA Notes, **16**: 78-84.

Cancian, A.K., 2001, *Reflexão e Colaboração Desencadeando Mudanças – Uma Experiência de Trabalho junto a Professores de Matemática*. Master thesis. Universidade Estadual Paulista. Rio Claro. Brazil.

Castells, M, 1999, *A sociedade em rede*, Editora Paz e Terra, São Paulo.

Castells, M., 2003, *A galáxia da internet: reflexões sobre a internet, os negócios e a sociedade*, Jorge Zahar Ed, Rio de Janeiro. Translated by: Borges, M.L. Translation of: Reflections on Internet, business and society.

Christiansen, I., 1997, When negotiation of meaning is also negotiation of task, *Educational Studies in Mathematics*. **34**(1): 1-25.

Clements, K., 1981, Visual imagery and school Mathematics (2nd. part), *For the learning of Mathematics*. **2**(2): 33-39.

Cobb, P., and Steffe, L., 1983, The constructivist Researcher as Teacher and Model Builder, *Journal for Research in Mathematics Education*. **14**(2): 83-94.

Confrey, J., 1991a, The Concept of Exponential Functions: A Student's Perspective, in: *Epistemological Foundations of Mathematical Experience*, L. Steffe, ed., Springer Verlag, New York, pp. 124-159.

Confrey, J., 1991b, *Function Probe* (software).Santa Barbara, CA, U.S.A.

Confrey, J., 1994, Voice and perspective: hearing epistemological innovations students' words, in: *Reveu des Sciences de l'education*. Special issue: Constructivism in Education, N. Bednarz, M. Larochelle, and J. Desautels, eds., **20**(1): 115-133.

Confrey, J., and Smith, E., 1994, Comments on James Kaput's Chapter, in: *Mathematical Thinking and Problem Solving*, A. H. Shoenfeld, ed., Lawrence Erlbaum Associates, New Jersey, pp. 172-192.

CRiME - Centre for Research in Mathematics Education, 2002, *What is experimental mathematics?* (October 28, 2002); http://www.soton.ac.uk/~crime/research/expmath/what_is_exp_math.html

Cruz, I, Presmeg, N., and Güemes, M., 2001, Reflections from two case studies of imagery and meaning in eighth grade mathematics, *Focus on Learning Problems in Mathematics*. 23(1): 1-16.

Cunningham, D., 1998, Cognition as semiosis: the role of inference, *Theory and Psychology*. 8: 827-840.

D'Ambrosio, U., 1978, Relationship of integrated science to other subjects in the curriculum. A paper presented at WG W-8, International Conference on Integrated Science Education Worldwide, Nijmegen, The Netherlands, April.

D'Ambrosio, U., 1985, Ethnomathematics and its place in the history and pedagogy of mathematics, *For the Learning of Mathematics*. **5**(1): 41-48.

D'Ambrosio, U., 2001, *Etnomatemática - Elo entre as Tradições e a Modernidade*, Editora Autêntica, Belo Horizonte.

Da Silva, H., 2000, *A Informática em Aulas de Matemática: a Visão das Mães*. Master thesis. Universidade Estadual Paulista. Rio Claro. Brazil.

Da Silva, M., 1999, *O Computador na Formação Inicial do Professor de Matemática: um Estudo a partir das Perspectivas de Alunos-Professores*. Master thesis. Universidade Estadual Paulista. Rio Claro. Brazil.

Davis, G., and Jones, K., 1996, Psychology of Experimental Mathematics, in: *Proceedings of the PME 20*, L. Puig, and A. Gutiérrez, A., eds., 1: 148.

Davis, P., 1993, Visual theorems, *Educational Studies in Mathematics*. 24(4): 333-344.

Davis, P., and Hersh. R., 1985, *A experiência matemática*, Francisco Alves Editora, Rio De Janeiro. Translated by J.B. Pitombeira.

Denzin, N., and Lincoln, Y. (Eds.), 2000, *Handbook of Qualitative Research*, Sage Publications.

Devaney, R., 1990, *Chaos, Fractals, and Dynamics. Computer experiments in mathematics*, Addison-Wesley Publishing Company.

Devlin, K., 1997, The logical structure of computer-aided mathematical reasoning, *American Mathematical Monthly*. **104**(7): 632-646.

Domingues, H., 2002, A demonstração ao longo dos séculos. *Boletim de Educação Matemática - BOLEMA*, **18**: 55-67.

Dreyfus, T., 1991, On the status of visual reasoning in mathematics and mathematics education, in: *Proceedings of PME 15*, F. Furinguetti, ed., **1**: 33-48.

Dubinsky, E., and Tall, D., 1991, Advanced mathematical thinking and the computer, in: *Advanced Mathematical Thinking*, D. Tall, ed., Kluwer Academic Publishers, Dordrecht, pp. 231-248.

Eisenberg, T., and Dreyfus, T., 1989, Spatial visualization in Mathematics curriculum, *Focus on learning problems in Mathematics*. **11**(1): 1-5.

Eisenberg, T., and Dreyfus, T., 1991, On the reluctance to visualize in Mathematics, in: *Visualization in teaching and learning Mathematics*, W. Zimmermann, and S. Cunningham, eds., Mathematical Association of America, Washington, pp. 25-38.

English, L., 1998, Children's problem posing within formal and informal contexts, *Journal for Research in Mathematics Education*. **29**(1): 83-106.

Epstein, D., Levy, S., and de la Llave, R., 1992, About this journal, *Experimental Mathematics*. **1**(1): 1-3.

Ernest, P., 1991, *The Philosophy of Mathematics Education*, The Falmer Press, Hampshire.

Etcheverry, N., Evangelista, N., Reid, M., Torroba, E., and Villarreal, M., in press, Fomentando discusiones en un ambiente computacional a través de la experimentación y la visualización. *Zetetiké*. Universide Estadual de Campinas, Faculdade de Educação, Círculo de Estudo, Memória e pesquisa em Educação Matemática.

Fiorentini, D., 1995, Alguns modos de ver e conceber o ensino da matemática no Brasil, *Zetetiké*. **3**(4): 1-39.

FOLDOP-Free on-line Dictionary of Philosophy, 2002, (December 3, 2002); http:/www.swif.uniba.it/lei/foldop/

Fontana, A., and Frey, J, 1994, Interviewing: the art of science, in: *Handbook of Qualitative Research*, N. Denzin, and Y. Lincoln., eds., Sage Publications, California, pp. 361-376.

Fontana, A., and Frey, J. , 2000, The interview: from structured questions to negotiated text, in: *Handbook of Qualitative Research*, N. Denzin, and Y. Lincoln., eds., Sage Publications, California, pp. 645- 672.

Francis, G., 1996, *Mathematical Visualization: Standing at the Crossroads*. Conference Panel for IEEE Visualization' 96, (October 28, 2002); http://www.cecm.sfu.ca/projects/PhilVisMath/vis96panel.html

Freire, P., 1976, *Educação como Prática da Liberdade*, Editora Paz e Terra, Rio de Janeiro.

Freire, P., 1981, *Pedagogia do Oprimido*, Editora Paz e Terra, Rio de Janeiro.

Freire, P., 1992, *Pedagogia da Esperança: um Reencontro com a Pedagogia do Oprimido*, Paz e Terra, Rio de Janeiro.

Garnica, V., 1992, *A Interpretação e o Fazer do Professor: a Possibilidade do Trabalho Hermenêutico na Educação Matemática*. Master dissertation. Universidade Estadual Paulista. Rio Claro. Brazil.

Garnica, V., 2002, As demonstrações em Educação Matemática: um ensaio, *Boletim de Educação Matemática - BOLEMA*. **18**: 91-99.

Gazetta, M., 1989, *A Modelagem como Estratégia de Aprendizagem na Matemática em Cursos de Aperfeiçoamento de Professores*. Master thesis. Universidade Estadual Paulista. Rio Claro. Brazil.

Gazire, E., 1988, *Perspectivas da Resolução de Problemas em Educação Matemática*. Master thesis. Universidade Estadual Paulista. Rio Claro. Brazil.

Giraldo, V., Carvalho, L., and Tall, D., 2003, Descriptions and definitions in the teaching of elementary calculus, in: in: *Proceedings of the PME 27 and PME-NA 25*, N. Pateman, B. Dougherty, and J. Zilliox, eds, **2**: 445-452.

Good, C. (Ed.), 1959, *Dictionary of Education*, Mc Graw-Hill Book Company, Inc.

Goody, J., 1977, *The domestication of the savage mind*, Cambridge University Press.

222

Gracias, T. S., 2003, *A Natureza da Reorganização do Pensamento em um Curso a Distância sobre "Tendências em Educação Matemática"*. Doctoral dissertation. Universidade Estadual Paulista. Rio Claro. Brazil.

Gustineli, O., 1990, *Modelagem Matemática e Resolução de Problemas: uma Visão Global em Educação Matemática*. Master thesis. Universidade Estadual Paulista. Rio Claro. Brazil.

Gutiérrez, A., 1996, Visualization in 3-dimensional geometry: in search of a framework, in: *Proceedings of the 20 PME*, L. Puig, and A. Gutiérrez, eds., 1: 3-20.

Gvirtz, S., 1999, *El discurso escolar a través de los cuadernos de clase. Argentina 1930-1970*, Buenos Aires, EUDEBA.

Habre, S., 2001, Visualization in Multivariable Calculus: the case of 3D-surfaces, *Focus on Learning Problems in Mathematics*. 23(1): 30-48.

Hanna, G., 2000, Proof, Explanation and Exploration: an Overview, *Educational Studies in Mathematics*. 44(1-2): 5-23.

Hanson, A, 1996, *Mathematical Visualization: Standing at the Crossroads*. Conference Panel for IEEE Visualization' 96 (October 28, 2002); http://www.cecm.sfu.ca/projects/PhilVisMath/vis96panel.html

Hébrard, J., 2001, Por uma bibliografia material das escritas ordinárias: o espaço gráfico do caderno escolar (França - Séculos XIX e XX), *Revista Brasileira de História da Educação*. 1(1): 115-141.

Hersh, R., 1997, *What is mathematics, really?*, Oxford University Press.

Heuvel-Panhuizen, M. (Ed.), 2001, *Proceedings of the 25th International Conference for the Psychology of Mathematics Education*. Volumes 1, 2, 3 e 4.

Horgan, J., 1993, La muerte de la demostración, *Investigación y Ciencia* (Spanish version of *Scientific American*). 207: 70-77.

Japiassú, H., and Marcondes, D., 1996, *Dicionário Básico de Filosofia*, Jorge Zahar Editor, Rio de Janeiro.

Jörgengon, L., 1996, *Mathematical Visualization: Standing at the Crossroads*. Conference Panel for IEEE Visualization' 96 (October 28, 2002); http://www.cecm.sfu.ca/projects/PhilVisMath/vis96panel.html

Kaiser-Messmer, G., 1991, Application-oriented mathematics teaching: a survey of the theoretical debate, in: *Teaching of mathematical modelling and applications*, M. Niss, W. Blum, and I. Huntley, I., eds., Ellis Horwood, pp. 83-92.

Kaput, J., 1994, Democratizing access to Calculus, in: *Mathematical Thinking and Problem Solving*, A. Schoenfeld, ed., Lawrence Erlbaum Associates, Inc., Publishers, Hillsdale, pp. 77-156.

Kenski, V., 2003, *Tecnologias e Ensino Presencial e a Distância*, Editora Papirus, Campinas.

Kerckhove, D., 1997, *A pele da cultura. (Uma investigação sobre a nova realidade eletrônica)*, Relógio D'Água Editores, Lisboa. Translated by Soares, L., and Carvalho, C. Translation of: The skin of culture (Investigating the new electronic reality).

Kilpatrick, J., 1987, Problem formulating: where do god problems come from?, in: *Cognitive Science and Mathematics Education*, A. Schoenfeld, ed., Lawrence Erlbaum Associates, Inc. Publishers, pp. 123-147.

Kline, M., 1970, Logic versus Pedagogy, *American Mathematical Monthly*. 77: 264-282.

Lakatos, I., 1976, *Proofs and Refutations: the logic of mathematical discovery*, Cambridge University Press, Cambridge.

Lakatos, I., 1978, *Mathematics, science, and epistemology*. Cambridge University Press, Cambridge.

Lalande, A., 1996, *Vocabulário técnico e crítico da Filosofia*, Livraria Martins Fontes Editora Ltda., São Paulo.

Lang, S., 1969, *Analysis I*, Addison-Wesley Publishing Company, INC.

Larson, R., Hostetler, R and Edwards, B., 1997, *College Algebra: A Graphing Approach*, Houghton Mifflin Company.

Latour, B., 1987, Science in action. How to follow scientists and engineers through society, Harvard University Press, Cambridge.

Lave, J., 1988, Cognition in practice. Mind, mathematics and culture in everyday life, Cambridge University Press, Cambridge.

Lean, G., and Clements, K., 1981, Spatial ability,visual imagery, and mathematical performance, *Educational Studies in Mathematics*. **12**(3): 267-299.

Lévy, P., 1993, *As tecnologias da inteligência. O futuro do pensamento na era da informática*, Editora 34, São Paulo. Translated by Costa, C. Translation of: Les technologies de l'intelligence.

Lévy, P., 1998, *A inteligência coletiva. Por uma antropologia do ciberespaço*, Edições Loyola, São Paulo. Translated by L. P. Rouanet. Translation of L'intelligence collective. Pour une antropologie du cyberspace.

Lévy, P., 1999, *Cibercultura*, Editora 34, São Paulo. Translated by C. I. da Costa. Translation of Cyberculture.

Lima, E., 1976, *Curso de Análise. Volume 1*, Livros Técnicos e Científicos Editora, Rio de Janeiro.

Lincoln, Y., and Guba, E., 1985, *Naturalistic Inquiry*, SAGE Publications, California.

Lourenço, M., 2002, A demonstração com informática aplicada à Educação, *Boletim de Educação Matemática – BOLEMA*.**18**: 100-111.

Mac Lane, S., 1996, Abstract Algebra uses homomorphisms, *American Mathematical Monthly*. **103**(4): 330-331.

Machado, N., *Ensaios Transversais: Cidadania e Educação*, Escrituras Editora, São Paulo.

Malheiros, A., 2004, *A Produção Matemática dos Alunos em um Ambiente de Modelagem*. Master thesis. Universidade Estadual Paulista. Rio Claro. Brazil.

Maltempi, M.V., 2003, Educação a Distância, *Revista Interface, Comunicação, Saúde, Educação*. **7**: 146.

Meira, L., 1998, Making sense of instructional devices: the emergence of transparency in mathematical activity. *Journal for Research in Mathematics Education*, 29 (2), 121-142.

Mellin-Olsen's, S., 1987, *The Politics of Mathematics Education*, D. Reidel Publishing Company, Dordrecht.

Meloy, T., and Barros, A., 2000, Base x Height: the transformation of a rectangle, *Hands On!*. **23**(2): 4-7.

Monagham, J., Sun, S., and Tall, D., 1994, Construction of the limit concept with a Computer Algebra System, in: *Proceedings of the 18 PME*, J. P. Ponte and J. F. Matos, eds., **3**: 279-286.

Moran, J.M., 2003, Contribuições para uma pedagogia da educação a distância no ensino superior, *Revista Interface, Comunicação, Saúde, Educação*. **7**: 147.

Moreira, P., and David, M.:2003, Matemática escolar, matemática científica, saber docente e formação de professores, *Zetetiké*. **11**(19): 57-80.

Nemirovsky R., and Noble, T., 1997, On mathematical visualization and the place where we live, *Educational Studies in Mathematics*. **33**(2): 99-131.

Niss, M., 1977, The "crisis" in mathematics instruction and a new teacher education at grammar school level, *International Journal of Mathematics Education in Science and Technology*. **8**(3): 303-321.

224

Noss, R. Healy, L., and Hoyles, C., 1997, The construction of mathematical meanings: connecting the visual with the symbolic, *Educational Studies in Mathematics*. **33**(2): 203-233.

Noss, R., and Hoyles, C., 1996, *Windows on mathematical meanings: learning cultures and computers*, Kluwer Academic Publishers, Dordrecht.

Oliver, A., and Newstead, K. (Ed.), 1998, *Proceedings of the 22nd International Conference for the Psychology of Mathematics Education*. Volumes 1, 2, 3 e 4.

Pateman, N., Dougherty, B., and Zilliox, J. (Eds.), 2003, *Proceedings of PME 27 and PME-NA 25*.

Patterson, N., and Norwood, K., 2004, A case study of teachers beliefs on students' beliefs about multiple representation, *International Journal of Science and Mathematics Education*. **2**(1): 5-23.

Penteado Silva, M., 1997, *O Computador na Perspectiva do Desenvolvimento Profissional do Professor*. Doctoral dissertation. Universidade Estadual de Campinas. Campinas. Brazil.

Penteado, M., 2001, Computes-based learning environments: risks and uncertainties for teachers, *Ways of Knowing Journal*. **I**(2): 23-35.

Penteado, M., and Borba, M. (Ed.), 2000, *A Informática em Ação - Formação de Professores, Pesquisa e Extensão*, Editora Olho d'Água, São Paulo.

Pollak, H., 1987, Cognitive Science and Mathematics Education: a mathematician's perspective, in: *Cognitive Science and Mathematics Education*, A. Schoenfeld, ed., Lawrence Erlbaum Associates, Inc. Publishers, pp. 253-264.

Polya, G., 1945, *How to Solve it?*, Princeton University Press.

Powell, A., and Ramnauth, M., 1992, Beyond questions and answers: prompting reflections and deepening understandings of mathematics using multiple-entry Logs, *For the learning of mathematics*. **12**(2): 12-18.

Presmeg, N., 1986a, Visualisation in mathematics giftedness, *Educational Studies in Mathematics*. **17**(3): 297-311.

Presmeg, N., 1986b, Visualization in High School Mathematics, *For the Learning of Mathematics*. **6**(3): 42-46.

Roche, D., 1996, As práticas da escrita nas cidades francesas do século XVIII, in: *Práticas da leitura*, R. Chartier, ed., Editora Estação Liberdade, São Paulo, pp. 177-200. Translation of *Pratiques de la lectura*. Translated by Nascimento, C.

Schaff, A., 1990, *A sociedade informática. As conseqüências sociais da segunda revolução industrial*, Editora UNESP, São Paulo. Translation of *Wohin führt der Weg*.

Scheffer, N., 2001, *Sensores, Informática e o Corpo: a Noção de Movimento no Ensino Fundamental*. Doctoral dissertation. Universidade Estadual Paulista. Rio Claro. Brazil.

Scheffer, N., 2002, *Corpo - Tecnologias - Matemática: uma interação possível no ensino fundamental*, EdiFAPES, Erechim, RS.

Schoenfeld, A., 1987, What's all the fuss about metacognition?, in: *Cognitive Science and Mathematics Education*, A. Schoenfeld, ed., Lawrence Erlbaum Associates, Inc. Publishers, pp. 189-215.

Schoenfeld, A., 1992, Learning to think mathematically: problem solving, metacognition and sense making in mathematics, in: *Handbook of research in mathematics teaching and learning*, D. Grouws, ed., Simon & Schuster Macmillan, New York, pp. 334-370.

Schoenfeld, A., 1994, Reflections on doing and teaching mathematics, in: *Mathematical thinking and problem solving*, A. Schoenfeld, ed., Lawrence Erlbaum Associates, Inc., Publishers, pp. 53-70.

Scucuglia, R., 2002, Calculadoras gráficas: conjecturando um teorema a partir de um estudo investigativo de funções, in: *Anais do V Simpósio de Iniciação Científica*, UNESP - Rio Claro, pp. 89.

Shank, G., and Cunningham, D., 1996, Modeling the six models of Peircean abduction for educational purposes. Paper presented at the Annual Meeting of the Midwest AI and Cognitive Science Conference, Bloomington, IN.

Silver, E., 1994, On mathematical problem posing, *For the Learning of Mathematics*. **14**(1): 19-28.

Silver, E., and Burkett, M., 1994, *The posing of division problems by preservice elementary school teachers*. Paper presented at the annual meeting of the American Educational Research Association, New Orleans, LA.

Silver, E., and Cai, J., 1996, An analysis of arithmetic problem posing by middle school students, *Journal for Research in Mathematics Education*. **27**(5): 521-539.

Silver, E., Mamona, J., Leung, S., and Kenney, P., 1996, Posing mathematical problems: an exploratory study, *Journal for Research in Mathematics Education*. **27**(3): 293-309.

Skovsmose, O., 1994, *Towards a Philosophy od Critical Mathematics Education*, Kluwer Academic Publishers, Dordrecht.

Skovsmose, O., 2000, Cenários para investigação. *Boletim de Educação Matemática-BOLEMA*, **14**: 66-91.

Skovsmose, O., and Borba, M., 2004, *Research methodology and critical mathematics education*, in: *Researching the Socio-political Dimensions of Mathematics Education: Issues of Power in Theory and Methodology*, R. Zevenbergen, and P. Valero, eds., Kluwer Academic Publishers, Dordrech, pp. 207-226.

Souza, T.:1996, *Calculadoras Gráficas: uma Proposta Didático-Pedagógica para o tema Funções Quadrátiacs*. Master thesis. Universidade Estadual Paulista. Rio Claro. Brazil.

Souza, T., and Borba, M., 1998, Calculadoras gráficas e funções quadráticas, *Revista de Educação Matemática*. **6**(4): 27-32.

Souza, T., and Borba, M., 2000, Explorando possibilidades e potenciais limitações da calculadora gráfica, *Educação e Matemática*. **56**(1): 35-39.

Stanic, G., and Kilpatrick, J., 1989, Historical perspectives on problem solving in the mathematics curriculum, in: *The teaching and assesing of mathematical problem solving*, C. Randall, and E. Silver, eds., Lawrence Erlbaum Associates, pp. 1-22.

Steffe, L., and Thompson, P., 2000, Teaching Experiment Methodology: Underlying principles and essential elements, in: *Research Design in mathematics and science education*, R. Lesh, and A. E. Kelly, eds., Erlbaum, Hillsdale, pp. 267-307.

Stylianou, D., 2001, On the reluctance to visualize in mathematics: is the picture changing?, in: *Proceedings of the 25 PME*, M. Van Den Heuvel-Panhuizen, ed., **4**: 225-232.

Tabach, M., 1999, Emphasizing multiple representations in algebraic activities, in: *Proceedings of the 23 PME*, O. Zaslavsky, ed., **1**: 322.

Tall, D., 1991, Intuition and rigor: the role of visualization in the Calculus, in: *Visualization in teaching and learning Mathematics*, W. Zimmermann, and S. Cunningham S., eds., Mathematical Association of America, Washington, DC, pp. 105-119.

Tall, D., 1993, Real Mathematics, rational computers and complex people, in: *Proceedings of the Fifth Annual International Conference on Technology in College Mathematics Teaching*, Addison-Wesley, pp. 243-258.

Tall, D., 1996, Information technology and Mathematics Education: enthusiasms, possibilities and realities, in: *Proceedings of the 8th International Congress on Mathematical Education*, C. Alsina, J. Álvarez, M. Niss, A. Pérez, L. Rico, and A. Sfard, eds., SAEM Thales, Sevilla, pp. 65-82.

Tall, D., and Thomas, M., 1989, Versatile learning and the computer. *Focus on learning problems in Mathematics*, 11 (2), 117-126.

The New Lexicon Webster's Dictionary of the English Language, 1989, Lexicon Publications, INC, New York.

Thomas, G., and Finney, R., 1979, *Calculus and Analytic Geometry*, Addison-Wesley Publishing Company, 5th ed.

Thurston, W., 1990, Mathematics Education, *Notices of the American Mathematical Society*. **37**: 844-850.

Thurston, W., 1994, On Proof and Progress in Mathematics, *Bulletin of the American Mathematical Society*. **30**(2): 161-177.

Thurston, W., 1995, Proof and progress in mathematics, *For the Learning of Mathematics*. **15**(1): 29-37.

Tikhomirov, O. K., 1981, The psychological consequences of computarization, in: *The Concept of Activity in Soviet Psychology*, J. V. Wertsch, ed., M.E. Sharpe Inc., New York, pp. 256-278.

Valente, J. A., 2003, Educação a distância no ensino superior: soluções e flexibilizações, *Revista Interface, Comunicação, Saúde, Educação*. **7**: 139-142.

Villarreal, M., 1999, *O Pensamento Matemático de Estudantes Universitários de Cálculo e Tecnologias Informáticas*. Doctoral dissertation. Universidade Estadual Paulista. Rio Claro. Brazil.

Villarreal, M., 2000, Mathematical thinking and intellectual technologies: the visual and the algebraic, *For the learning of Mathematics*. **20**(2): 2-7.

Villarreal, M and Borba, M., 1996, Computers and Calculus: visualization and experimantation to characterize extremes of functions, in: *Book of abstracts of short presentations ICME 8*, Sevilla, pp. 408.

Vithal, R., Christiansen, I., and Skovsmose, O., 1995, Project work in university mathematics education, *Educational Studies in Mathematics*. **29**(2): 199-223.

Wagner, H., 1979, *Fenomenologia e relações sociais - Textos escolhidos de Alfred Schutz*, Zahar Editores, Rio de Janeiro. Translated by Melin, A. Translation of *Alfred Schutz on Phenomenology and Social Relations*.

Webster's Third New International Dictionary of the English Language,1966, G and C. Merriam Co.

Zanin, A. C., 1997, *O LOGO na Sala de Aula de Matemática da 6ta série do 1º grau*. Master thesis. Universidade Estadual Paulista. Rio Claro. Brazil.

Zaslavsky, O. (Ed.), 1999, *Proceedings of the 23rd International Conference for the Psychology of Mathematics Education*. Volumes 1, 2, 3 and 4.

Zazkis, R., Dubinsky, E., and Dautermann, J., 1996, Coordinating visual and analytic strategies: a study of students' understanding of the group D4, *Journal for Research in Mathematics Education*. **27**(4): 435-457.

Zimmermann, W., and Cunningham, S., 1991, Editors' Introduction: What is Mathematical Visualization?, in: *Visualization in teaching and learning Mathematics*, W. Zimmermann, and S. Cunningham S., eds., Mathematical Association of America, Washington, DC, pp. 1-8.

Index

Alrø 120, 205
Araújo 58, 120, 193

Barbosa 54, 108, 121, 180, 193
Bassanezi 36, 37, 44, 46, 49, 50, 51, 52, 60, 116
Benedetti 125, 135, 189, 191
Bicudo, I. 66
Bicudo, M.A.V. 32, 188
Bishop 89
Blum 36

Cancian 195
Castells 170, 203, 204
CBR 2, 77, 140, 143, 144, 165, 169, 191

Chat 172, 173, 178, 179, 181, 184, 185
Christiansen 52, 53, 54

Citizenship 45, 50, 57, 59, 60, 101, 204, 205, 207, 214, 215
Clements 80
Cobb 16, 125, 189
Confrey 15, 16, 73, 76, 78, 93, 97, 163, 189, 197
Conic sections 130
Cunningham, D. 75, 122
Cunningham, S. 80, 81, 90, 157

D'Ambrosio 26, 48, 49, 50, 52, 54, 60, 188, 205, 207
Da Silva 193, 195, 198
Dautermann 81, 82, 166
Davis, G. 71
Davis, P. 35, 86
Davis,. P. 66
Denzin 188

Derivative 74, 105, 107, 119, 126, 145, 149, 150
Dichotomy internal-external 82, 166
Distance education 170, 171, 172, 179, 184, 185, 195, 207
Domingues 87
Dreyfus 80, 81, 83, 86, 88, 89, 94, 95
Dubinsky 81, 82, 90, 92, 98, 166

Eisenberg 80, 81, 89, 94, 95
English 22, 40, 52, 121, 138, 173
Etcheverry 130
Ethnomathematics 26, 27, 36, 37, 49, 121, 188, 205
Experimental-with-technology 64, 75, 115, 165, 187
Experimentation examples
    Body movement 141
    Conic sections 130
    Derivative 145
    Functions 135, 140
    Parabola 126
    Tangent lines 151
    with computer 130, 135
    with graphing calculator 140

Fiorentini 188
Fontana 191
Freire 17, 48, 50, 52, 54, 60, 180, 188
Frey 191
Functions 64, 73, 109, 113, 119, 126, 135, 140, 153, 155, 181

Garnica 17, 87
Gazetta 116, 118
Gazire 34, 35
Giraldo 155

GPIMEM 4, 71, 73, 104, 113, 120, 140, 171, 179, 187, 189, 196, 199, 205
Gracias 129, 172, 173, 185, 195
graphing calculator 2, 64, 72, 74, 76, 102, 113, 128, 129, 140, 141, 143, 144, 145, 156, 166, 192, 198
Guba 29, 30, 32, 54, 184, 185, 188, 197, 199
Gustineli 44
Gutiérrez 81
Gvirtz 202, 203

Hanna 83, 86
Healy 72
Hersh 35, 66
Horgan 87
Hoyles 15, 20, 72
Humans-with-media 23, 27, 32, 56, 73, 78, 92, 98, 104, 108, 122, 158, 164, 165, 167, 178, 189, 191, 197, 198, 201, 205, 208, 211
Humans-with-textbooks 105

Intermedia coordination 144, 154, 167, 190
Intershaping relationship 15, 16, 17, 20, 22, 98, 166, 189, 191

Kaput 120, 163, 197
Kenski 19
Kerckhove 24, 25, 26, 122
Kilpatrick 33, 34, 35, 38, 43, 46

Lakatos 67, 75, 212
Lave 12, 20, 21
LBM 190
Lean 80
Lévy 17, 20, 21, 22, 23, 24, 25, 26, 27, 72, 82, 87, 92, 114, 121, 158, 166, 173, 178, 185, 190, 191, 201, 202, 204, 208
Lincoln 29, 30, 32, 54, 184, 185, 188, 197, 199
Lourenço 73

Machado 10
Malheiros 103, 108, 113, 192
Mean Value Theorem 118, 119, 159, 160, 162, 163, 164

Modeling
    and demonstration 114
    and Ethnomathematics 49
    and problem posing 42
    and problem solving 36
    and project work 48
    and technology 54, 58
    and the Internet 109
    as a pedagogical approach 29, 102
    Limitations 59
    Roots in Brazil 47
Modeling examples
    chloroplasts 105
    mad cow 110
    potatoes plantation 116
Multiple representation 16, 78

Nemirovsky 72, 78, 81, 82, 96, 98, 207
Niss 52, 53, 60
Noss 15, 20, 72
Notebook 201, 202

Penteado 29, 108, 129, 172, 173, 179, 194, 195
Political aspects 3, 45, 101, 173, 188, 201, 203, 206, 211
Pollak 43
Polya 33, 34, 35, 66, 67
potatoes 44, 116, 117
Powell 73, 207
Presmeg 80, 81, 95, 144, 153
Problem posing 37
Problem solving 33
Project work 52

Reorganization of thinking 3, 5, 14, 27, 78, 92, 101, 122, 125, 165, 166, 167, 169, 173, 187, 195
Research group 4, 27, 63, 189, 207
research methodology 5, 32, 184, 193, 194, 196, 207, 209

Schaff 9, 10
Scheffer 77, 135, 140, 141, 165, 190, 191
Schoenfeld 33, 35, 66, 67
Scucuglia 128
Silver 38, 39, 40, 41, 42, 43
Skovsmose 52, 53, 54, 58, 59, 120, 129, 193, 195, 205, 207, 215

Souza 73, 77, 190, 191
Steffe 16, 125, 189

Tall 90, 91, 92, 98, 155
Tangent 133
Tangent lines 108, 145, 149, 151
Teaching experiment 130, 140, 145, 150,
    189, 191, 192
TERC 78, 190, 207
Thinking collective 82, 92, 96, 97, 139,
    150, 165, 187, 190, 191, 198
Thompson 125, 189
Thurston 83, 115, 139, 163

Tikhomirov 11, 12, 13, 14, 15, 20, 27, 92,
    190

Videopaper 77
Visualization
    and proofs 158
    and textbooks 158
    definitions 79
Vithal 52, 53

Zanin 192
Zazkis 81, 82, 166
Zimmermann 80, 81, 90, 157

Mathematics Education Library

Managing Editor: A.J. Bishop, Melbourne, Australia

1. H. Freudenthal: Didactical Phenomenology of Mathematical Structures. 1983
   ISBN 90-277-1535-1 HB; 90-277-2261-7 PB
2. B. Christiansen, A. G. Howson and M. Otte (eds.): Perspectives on Mathematics
   Education. Papers submitted by Members of the Bacomet Group. 1986
   ISBN 90-277-1929-2 HB; 90-277-2118-1 PB
3. A. Treffers: Three Dimensions. A Model of Goal and Theory Description in
   Mathematics Instruction TheWiskobas Project. 1987   ISBN 90-277-2165-3
4. S. Mellin-Olsen: The Politics of Mathematics Education. 1987
   ISBN 90-277-2350-8
5. E. Fischbein: Intuition in Science and Mathematics. An Educational Approach.
   1987   ISBN 90-277-2506-3
6. A.J. Bishop: Mathematical Enculturation. A Cultural Perspective on
   Mathematics Education.1988 ISBN 90-277-2646-9 HB; 1991 0-7923-1270-8 PB
7. E. von Glasersfeld (ed.): Radical Constructivism in Mathematics Education.
   1991   ISBN 0-7923-1257-0
8. L. Streefland: Fractions in Realistic Mathematics Education. A Paradigm of
   Developmental Research. 1991   ISBN 0-7923-1282-1
9. H. Freudenthal: Revisiting Mathematics Education. China Lectures. 1991
   ISBN 0-7923-1299-6
10. A.J. Bishop, S. Mellin-Olsen and J. van Dormolen (eds.): Mathematical
    Knowledge: Its Growth Through Teaching. 1991   ISBN 0-7923-1344-5
11. D. Tall (ed.): Advanced Mathematical Thinking. 1991   ISBN 0-7923-1456-5
12. R. Kapadia and M. Borovcnik (eds.): Chance Encounters: Probability in
    Education. 1991   ISBN 0-7923-1474-3
13. R. Biehler, R.W. Scholz, R. Sträßer and B. Winkelmann (eds.): Didactics of
    Mathematics as a Scientific Discipline. 1994   ISBN 0-7923-2613-X
14. S. Lerman (ed.): Cultural Perspectives on the Mathematics Classroom. 1994
    ISBN 0-7923-2931-7
15. O. Skovsmose: Towards a Philosophy of Critical Mathematics Education. 1994
    ISBN 0-7923-2932-5
16. H. Mansfield, N.A. Pateman and N. Bednarz (eds.): Mathematics for
    Tomorrow's Young Children. International Perspectives on Curriculum.   1996
    ISBN 0-7923-3998-3
17. R. Noss and C. Hoyles:Windows on Mathematical Meanings. Learning Cultures
    and Computers. 1996   ISBN 0-7923-4073-6 HB; 0-7923-4074-4 PB
18. N. Bednarz, C. Kieran and L. Lee (eds.): Approaches to Algebra. Perspectives
    for Research and Teaching.1996   ISBN 0-7923-4145-7 HB; 0-7923-4168-6, PB
19. G. Brousseau: Theory of Didactical Situations in Mathematics. Didactique des
    Mathématiques 1970-1990. Edited and translated by N. Balacheff, M. Cooper,
    R. Sutherland and V. Warfield. 1997   ISBN 0-7923-4526-6
20. T. Brown: Mathematics Education and Language. Interpreting Hermeneutics
    and Post-Structuralism. 1997   ISBN 0-7923-4554-1 HB
    Second Revised Edition. 2001   ISBN 0-7923-6969-6 PB

Mathematics Education Library

21. D. Coben, J. O'Donoghue and G.E. FitzSimons (eds.): Perspectives on Adults Learning Mathematics. Research and Practice. 2000   ISBN 0-7923-6415-5
22. R. Sutherland, T. Rojano, A. Bell and R. Lins (eds.): Perspectives on School Algebra. 2000   ISBN 0-7923-6462-7
23. J.-L. Dorier (ed.): On the Teaching of Linear Algebra. 2000
      ISBN 0-7923-6539-9
24. A. Bessot and J. Ridgway (eds.): Education for Mathematics in the Workplace. 2000   ISBN 0-7923-6663-8
25. D. Clarke (ed.): Perspectives on Practice and Meaning in Mathematics and Science Classrooms. 2001   ISBN 0-7923-6938-6 HB; 0-7923-6939-4 PB
26. J. Adler: Teaching Mathematics in Multilingual Classrooms. 2001
      ISBN 0-7923-7079-1 HB; 0-7923-7080-5 PB
27. G. de Abreu, A.J. Bishop and N.C. Presmeg (eds.): Transitions Between Contexts of Mathematical Practices. 2001   ISBN 0-7923-7185-2
28. G.E. FitzSimons:What Counts as Mathematics? Technologies of Power in Adult and Vocational Education. 2002   ISBN 1-4020-0668-3
29. H.Alrø andO. Skovsmose:Dialogue and Learning in Mathematics Education. Intention, Reflection,Critique.2002 ISBN 1-4020-0998-4 HB;1-4020-1927-0 PB
30. K. Gravemeijer, R. Lehrer, B. van Oers and L. Verschaffel (eds.): Symbolizing, Modeling and Tool Use in Mathematics Education. 2002   ISBN 1-4020-1032-X
31. G.C. Leder, E. Pehkonen and G. Törner (eds.): Beliefs: A Hidden Variable in Mathematics Education? 2002   ISBN 1-4020-1057-5 HB; 1-4020-1058-3 PB
32. R. Vithal: In Search of a Pedagogy of Conflict and Dialogue for Mathematics Education. 2003   ISBN 1-4020-1504-6
33. H.W. Heymann: Why Teach Mathematics? A Focus on General Education. 2003 Translated by Thomas LaPresti   ISBN 1-4020-1786-3
34. L. Burton: Mathematicians as Enquirers: Learning about Learning Mathematics. 2004   ISBN 1-4020-7853-6 HB; 1-4020-7859-5 PB
35. P. Valero, R. Zevenbergen (eds.): Researching the Socio-Political Dimensions of Mathematics Education: Issues of Power in Theory and Methodology. 2004
      ISBN 1-4020-7906-0
36. D. Guin, K. Ruthven, L. Trouche (eds.) The Didactical Challenge of Symbolic Calculators: Turning a Computational Device into a Mathematical Instrument 2005   ISBN 0-387-23158-7
37. J. Kilpatrick, C. Hoyles, O. Skovsmose (eds. in collaboration with Paola Valero) Meaning in Mathematics Education. 2005   ISBN 0-387-24039-X
38. H. Steinbring. The Construction of New Mathematical Knowledge in Classroom Interaction: An Epistemological Perspective. 2005.   ISBN 0-387-24251-1
39. M.Borba, M. Villarreal: Humans-with-Media and the Reorganization of Mathematical Thinking: Information and Communication Technologies, Modeling, Visualization and Experimentation. 2005   ISBN 0-387-24263-5